Joel Dorman Steele

Fourteen Weeks in Natural Philosophy

Joel Dorman Steele

Fourteen Weeks in Natural Philosophy

ISBN/EAN: 9783337030490

Printed in Europe, USA, Canada, Australia, Japan

Cover: Foto ©berggeist007 / pixelio.de

More available books at **www.hansebooks.com**

HIGH-PRESSURE STEAM ENGINE.

FOURTEEN WEEKS

IN

NATURAL PHILOSOPHY,

BY

J. DORMAN STEELE, Ph.D.,

AUTHOR OF THE FOURTEEN-WEEKS SERIES IN PHYSIOLOGY, CHEMISTRY,
ASTRONOMY, AND GEOLOGY.

"The works of God are fair for naught,
Unless our eyes, in seeing,
See hidden in the thing the thought
That animates its being." —Tilton.

A. S. BARNES & COMPANY,
NEW YORK, CHICAGO AND NEW ORLEANS.

THE FOURTEEN WEEKS' COURSES
IN
NATURAL SCIENCE,
BY
J. DORMAN STEELE, A.M., PH.D.

Fourteen Weeks in Natural Philosophy,
Fourteen Weeks in Chemistry, (2 Editions)
Fourteen Weeks in Descriptive Astronomy,
Fourteen Weeks in Popular Geology,
Fourteen Weeks in Human Physiology,
A Key, containing Answers to the Questions
and Problems in Steele's 14 Weeks' Courses,

A HISTORICAL SERIES,
on the plan of Steele's 14 Weeks in the Sciences,

A Brief History of the United States,
A Brief History of France,

The publishers of this volume will send either of the above by mail, post-paid, on receipt of the price.

The same publishers also offer the following standard scientific works, being more extended or difficult treatises than those of Prof. Steele, though still of Academic grade.

Peck's Ganot's Natural Philosophy,
Porter's Principles of Chemistry,
Jarvis' Physiology and Laws of Health,
Wood's Botanist and Florist,
Chambers' Elements of Zoology,
McIntyre's Astronomy and the Globes,
Page's Elements of Geology,

Address A. S. BARNES & CO.,
Educational Publishers,
NEW YORK OR CHICAGO.

ENTERED according to Act of Congress, in the year 1869, by
A. S. BARNES & CO.,
In the Clerk's Office of the District Court of the United States for the Southern District of New York.

STEELE'S NAT. PHIL.

TO

My Wife,

WHOSE

SYMPATHY AND ASSISTANCE

IN MY SCIENTIFIC STUDIES

HAVE BEEN

My Constant Inspiration and Support,

This Volume

IS

Affectionately Dedicated.

PREFACE.

THIS work has grown up in the class-room. It contains those definitions, illustrations, and applications which seemed at the time to interest and instruct the pupils. Whenever any explanation fixed the attention of the learner, it was laid aside for future use. Thus, by steady accretions, the process has gone on until a book is the result.

As Physics is generally the first branch of Natural Science pursued in schools, it is important that the beginner should not be disgusted with the abstractions of the subject. The Author has therefore endeavored to use such simple language and practical illustrations as will interest the learner, while he is at once led out into real life. From the multitude of philosophical principles, those only have been selected which are essential to the information of every well-read person. The many curious questions which yet rank only as "philosophical gossip," are rarely mentioned. Within the brief limits of a small text-book, no subject can be exhaustively treated. This is, however, of less importance now, when every teacher feels that he must of necessity be above and beyond any school-work in the fulness of his information.

The theories advanced are those generally received among scientific men. The object of an elementary work is not to advance the peculiar ideas of any one person, but simply to give such currently accepted facts as are believed by all. This plan affords no scope for original thought. The Author has therefore simply sought to gather from every attainable source the freshest and most valuable information, and so weave it together as to please as well as instruct his pupils. The time-honored classifications established by the masters, and recognized in all scientific works, have been retained. It has not seemed wise to reject familiar terms for a mere appearance of novelty. Since the problems are for the instruction of American youth, the system of weights in general use in this country is employed.

The object of the Author will be fully attained if he succeed in leading some young mind to become a lover and interpreter of Nature, and thus come at last to see that Nature herself is but a "thought of God."

PECK'S GANOT'S NATURAL PHILOSOPHY, from the French of A. Ganot. By Prof. W. G. Peck, LL.D., Columbia College. Adapted to follow this work, or take its place in classes requiring a fuller course.

"Peck's Ganot" is noted for its clear method and its magnificent system of illustration (from the original French plates), largely doing away with the necessity for Apparatus.

The work contains 500 12mo pages, and is of Academic grade. Price $1.75, post-paid.   A. S. BARNES & CO., *Publishers*.

SUGGESTIONS TO TEACHERS.

SCHOLARS are expected to obtain information from this book, without the aid of questions, as they must always do in their general reading. When the subject of a paragraph is announced, the pupil should be prepared to tell all he knows about it. The diagrams and illustrations, as far as possible, should be drawn upon the blackboard and explained. Although pupils may, at first, manifest an unwillingness to do this, yet in a little time it will become a most interesting feature of the recitation. The "Practical Questions" given at the close of each general subject have been found a profitable exercise in awakening inquiry and stimulating thought. They may be used at the pleasure of the instructor. The equations contained in the text are designed to be employed in the solution of the problems. The following works will be found useful in furnishing additional illustrations and in elucidating difficult subjects, viz.: Deschanel's Natural Philosophy, Lockyer's Guillemin's Forces of Nature, Stewart's Elementary Physics, Herschel's Introduction to the Study of Physical Science, Tomlinson's Introduction to the Study of Natural Philosophy, Knight's Cyclopædia—Section on Science and Natural History, Pepper's Play-book of Science, Beale's How to Work with the Microscope, Schellen's Spectrum Analysis, Lockyer's The Spectroscope, Nugent's Optics, Chevreul on Colors, Thomson & Tait's Natural Philosophy, Maxwell's Electricity and Magnetism, Faraday's Forces of Matter, Youman's Correlation of Physical Forces, Maury's Physical Geography of the Sea, Atkinson's Ganot's Physics, Silliman's Physics, Tyndall's Lectures on Light, Heat and Sound, Tyndall's Forms of Water, Snell's Olmsted's Philosophy, Loomis's Meteorology, Miller's Chemical Physics, Cooke's Religion and Chemistry, Benedicite, and the Illustrated Library of Wonders. They may be procured of the publishers of this book. The author will be pleased to correspond with teachers with regard to the apparatus necessary for the performance of the experiments named in the work, or with reference to any of the "Practical Questions."

TABLE OF CONTENTS.

I. INTRODUCTION.

	PAGE
MATTER	13
General Properties	15
Specific Properties	27

II. ATTRACTION . . . 33

MOLECULAR FORCES	35
Cohesion	36
Adhesion	40
GRAVITATION	48
Weight	
Falling Bodies	50
Centre of Gravity	54
The Pendulum	58

III. MOTION . . . 68

LAWS OF MOTION	70
Compound Motion	71
Composition and Resolution of Forces	72
Circular Motion	75
Reflected Motion	79

IV. MECHANICAL POWERS . . 83

THE ELEMENTS OF MACHINERY	85
Lever	85
Wheel and Axle	90
Inclined Plane	93
Screw	95
Wedge	96
Pulley	97

V. PRESSURE OF LIQUIDS AND GASES . 101

HYDROSTATICS 103
   Liquids Influenced by Pressure . . . 103
   Liquids Influenced by Gravity alone . . 107
   Specific Gravity 115
HYDRAULICS 122
   Water-wheels 125
   Wave-motion 128
PNEUMATICS 132
   Properties of the Air 133
   Pressure of the Air, Barometer, Pumps, Siphon, &c. 136

VI. ON SOUND . . . 149

ACOUSTICS 151
   Sound Produced by Vibrations . . . 151
   Velocity of Sound 155
   Intensity. Speaking-tubes, Trumpet, &c. . 157
   Refraction 159
   Reflection. Echoes, &c. 160
   Noise and Music 163
   The Siren, Length of Sound-wave, &c. . 164
   Interference of Sound 168
   Vibration of Cords 169
   Nodes 171
   Wind Instruments 176
   The Ear 178
   Singing Flames 182

VII. ON LIGHT . . . 185

OPTICS 187
   Laws and Theory of Light . . . 188
   Reflection. Mirrors, &c. 190
   Refraction. Prisms, &c. 197
   Composition. The Spectrum, Diffraction, Polarization, the Rainbow, &c. . . 204

OPTICAL INSTRUMENTS. TELESCOPE, MICROSCOPE,
   OPERA-GLASS, MAGIC LANTERN, STEREOSCOPE,
   CAMERA, &c. 215
  THE EYE 220

VIII. ON HEAT . . . 225

NATURE AND THEORY OF HEAT . . . 228
CHANGE OF STATE BY HEAT . . . 232
  EXPANSION 233
  LIQUEFACTION 236
  VAPORIZATION 237
  EVAPORATION 241
COMMUNICATION OF HEAT . . . 243
  CONDUCTION 243
  CONVECTION 244
  RADIATION 245
THE STEAM-ENGINE 246
METEOROLOGY 248

IX. ON ELECTRICITY . . 259

MAGNETIC ELECTRICITY 261
FRICTIONAL ELECTRICITY 269
GALVANIC ELECTRICITY 287
ELECTRO-MAGNETISM 302
ANIMAL ELECTRICITY 314

CONCLUSION 314

NOTES ON APPARATUS AND EXPERIMENTS . 318
QUESTIONS 323
INDEX 339

NATURAL PHILOSOPHY.

INTRODUCTION.

MATTER is anything we can perceive with our senses. A *body* is a distinct portion of matter. Ex.: A chair, 1 lb. of iron, a piece of silver. A *substance* is any one of the different kinds of matter. Ex.: Gold, wood, stone.

PROPERTIES OF MATTER.—Each substance possesses two kinds of properties—*general* and *specific;* the former belongs to all substances, the latter only to particular ones. Ex.: Gold has weight. This property is not peculiar to gold, for all substances have weight; hence it is a *general* property. But gold is yellow. This is so distinctive that we speak of a "golden color;" hence it is a *specific* property. A piece of glass has form. All bodies have form; hence it is a *general* property. But glass is so brittle that we say "brittle as glass;" hence brittleness is a *specific* property.

CHANGES OF MATTER.—Each substance can undergo two kinds of change—*physical* and *chemical*. The former does not destroy the specific properties of the substance; the latter does. Ex.: An eagle is beaten into gold-leaf; but this does not alter the color or other specific properties of the metal, hence it is a *physical* change. Melt the eagle in a crucible, and still the same is true. But put it into an acid, and soon it is dissolved—the specific properties are entirely destroyed; hence it is a *chemical* change. Draw a nail into wire, and the specific properties of the iron remain untouched. But leave it in a basin of water, and the distinguishing properties of iron disappear—it becomes brittle, red, soft, and scaly; hence this is a *chemical* change.

These distinctions give rise to two branches of Natural Science, PHILOSOPHY and CHEMISTRY. The former treats of the physical, the latter of the chemical changes of matter. As all bodies possess both kinds of properties, and are susceptible of both kinds of change, these branches are intimately connected.

Practical Questions.—My knife-blade is magnetized, so that it will pick up a needle; is that a physical or chemical change? Is it treated in Philosophy or Chemistry?

Is the burning of coal a physical or chemical change? The production of steam? The formation of dew? The falling of a stone? The growth of a tree? The flying of a kite? The chopping of wood? The explosion of powder? The boiling of water? The melting of iron? The drying of clothes? The freezing of water?

The General Properties of Matter.

We cannot imagine a body which does not possess all the general properties of matter. The most important are Magnitude, Impenetrability, Divisibility, Porosity, Inertia, and Indestructibility.

Magnitude is the property of occupying space. Size is the amount of space a body fills. Every body has three dimensions—length, breadth, and thickness. In order to measure these, some standard is required. Anciently, certain portions of the human body were used for this purpose. Ex.: The foot; the cubit, or length of the forearm from the elbow to the end of the middle finger; the finger's length or breadth; the hand's breadth; the span, etc.

The English system.—The first intimation that is given of an attempt to have a standard in England, is that of 1120. King Henry ordered that the ell, the ancient yard, should be the exact length of his arm. Afterward we learn that a standard yardstick was kept at the Exchequer in London; but it was so inaccurate, that a commissioner, who examined it in 1742, wrote: "A kitchen poker filed at both ends would make as good a standard. It has been broken, and then repaired so clumsily that the joint is nearly as loose as a pair of tongs." In 1760, Mr. Bird prepared an accurate copy of this for the use of the Government. It was not legally adopted until 1824, when it was ordered, that if destroyed it should be restored by a comparison with the length of a pendulum vibrating seconds at the latitude of London. (See page 59, 4th Law.)

At the great fire in London, 1834, the Parliament House was burned, and with it Bird's yard-stick. Repeated attempts were then made to find the length of the lost standard by means of the pendulum. This was found utterly impracticable. At last the British Government adopted a standard prepared from the most reliable copies of Bird's yard-stick. A copy of this was taken by Troughton, a celebrated instrument-maker of London, for the use of our Coast Survey.*

The French Standard.—The French adopted as the length of the legal foot, that of the royal foot of King Louis XIV., as perishable a standard as King Henry's arm. In 1790, however, they took $\frac{1}{10,000,000}$ of the length of the quarter of a meridian of the earth's circumference as the basis of all measurements. This is equal to 39 + inches, and is called a metre.

A Natural Standard.—Two attempts have thus been made to fix upon something in Nature as an invariable unit of measure. The French had, however, scarcely completed their system, when they found that a mistake had been made in measuring

---

* This yard is about $\frac{1}{1000}$ of an inch longer than the British standard. According to Act of Congress, sets of weights and measures have been distributed to the governors of the several states. The yards so furnished are equal to that of the Troughton scale. We have no national standard established by law.

the meridian. The English philosophers discovered a similar error in the calculation of the length of the pendulum. Both the French and English systems are therefore founded upon arbitrary standards.

IMPENETRABILITY is the property of so occupying space as to exclude all other bodies. No two bodies can occupy the same space at the same time. A book lies upon the table before me; no power in the world is able to place another in the same space with that. I attempt to fill a bottle through a closely-fitting funnel; but before the liquid can run in, the air must gurgle out or the water will trickle down the sides of the bottle.

Apparent Exceptions.—In common language, we speak of one substance penetrating another. Thus, a needle penetrates cloth, a nail penetrates wood, etc.; but a moment's examination shows that they merely push aside the fibres of the cloth or wood, and so press them closer together. Take a tumbler brim-full of water, and then cautiously drop in shingle-nails. A quarter of a pound can be easily added without causing the water to overflow. We shall find the explanation of this, in the fact that the surface of the water becomes convex.

DIVISIBILITY is that property of a body which allows it to be separated into parts. We have never seen a particle so small that we could not make it smaller.

Illustrations.—The thread with which certain species of spiders weave their web is composed of four smaller threads; each one of these consists of four smaller, each of which comes from a separate tube in the spider's spinning-machine. A German naturalist, after examining a web very carefully, decided that it would take 4,000,000,000 fibres to form a thread as large as a hair of his beard; and as each fibre is composed of 4,000 smaller ones, it follows that each of the least fibres is only $\frac{1}{16,000,000,000,000}$ part the size of a human hair. It is said that a half-pound of the full-sized thread would girt the globe. It would require 50,000,000 pounds of wire to erect a telegraph around the earth at the equator. A grain of strychnine will impart a flavor to 1,750,000 grains of water; hence, each grain of the liquid will contain only $\frac{1}{1,750,000}$ of a grain of strychnine, and yet that amount can be distinctly tasted. A grain of Magenta will color 50,000,000 grains of water. A piece of silver containing only one-billionth of a cubic inch—*i. e.*, being .001 of an inch square—when dissolved in nitric acid will render milky a solution of a hundred cubic inches of common salt. Each cubic inch must then contain $\frac{1}{100,000,000,000}$ of a cubic inch of silver. The eye can see the color distinctly in, perhaps, a hundredth part of a cubic inch, in which there would be present only one ten-trillionth part of a cubic inch of silver.*

---

* Some idea of the vastness expressed by the word trillion may be derived from the following curious calculation. If Adam, at the instant of his creation, had commenced to count at the

Press a puff-ball, and each speck of the cloud which flies off, under the microscope proves to be a beautiful round orange-ball,—the seed of the plant. Much of the fine dust that is revealed to us in the atmosphere, by a beam of light shining through a crevice, consists of the seeds of minute plants, which falling on a damp surface grow into mildew or mould. Under a microscope, this becomes a fairy forest of trees of a new and strange growth.

In all these instances we have mentioned, the divisibility is proved by our senses of taste and sight. When our eyes fail, the microscope is called in to continue the investigation. While thus, practically, there is no limit to the divisibility of matter, philosophers hold that there is, in theory.

THE ATOMIC THEORY supposes that matter is composed of inconceivably minute portions called atoms, each having a definite shape, weight, color, etc., which cannot be changed by any chemical or physical force. As has been happily said, "What God made one in the beginning, man cannot put asunder." No one has ever seen one of these ultimate portions of matter, and we have no absolute proof that any exist; but the theory is so conve-

---

rate of one every second of time, continuing through all the centuries, he would not yet have nearly completed the first quarter of a trillion; and even if Eve had come to his relief, and together they had counted day and night, they would not see the end of their task and enjoy their first leisure for 10,000 years to come.

nient, especially in chemistry, that it is at present generally received.

ANIMALCULÆ.—The tiny nations of animalculæ furnish most striking illustrations of the divisibility of matter and the minuteness of atoms. This is a world of which our unaided senses furnish us no proof. The microscope alone reveals its wonders. In the drop of water that clings to the point of a cambric needle, the swarming millions of this miniature world live, grow, and die. They swim in this their ocean, full of life, frisking, preying upon each other, waging war, and re-enacting the scenes of the great world we see about us. Myriads of them inhabit the pools of water standing along the roadside in summer. They go up in vapor and fly off in dust, and reappear wherever moisture and heat favor the development of life. Yet, minute as they are, they have been fossilized (turned to stone), and now form masses of chalk. Tripoli, or polishing-slate, is composed of these remains, each skeleton weighing the $\frac{1}{187,000,000}$ of a grain. If we examine whiting under a powerful microscope, we shall find that it is composed of *tiny shells*. Now let our imagination conceive the minute animals which formerly occupied them. Many of them had simple sack-like bodies, but still they had one or more stomachs, and possessed the power of digesting and assimilating food. This food, coursing in infinitely minute channels, must have been composed of solid as well as liquid matter; and finally, at the lowest extreme of this descending series, we come to the atoms of which this matter itself was composed.

POROSITY is the property of having pores. By this is meant not only such pores as are familiar to us all, and to which we refer when in common language we speak of a porous body, as bread, wood, unglazed pottery, a sponge, etc., but a finer kind, which are as invisible to the eye as the atoms themselves. These pores are caused by the fact that the molecules* of which a body is composed, are not in actual contact, but are separated by extremely minute spaces.

Size of the spaces compared with the size of the atoms. —These spaces are so small that they cannot be discerned with the most powerful microscope, yet it is thought that they are very large as compared with the size of the atoms themselves. If we imagine a being small enough to live on one of the atoms near the centre of a stone, as we live on the earth, then we are to suppose that he would see the nearest atoms at great distances from him, as we see the moon and stars, and might perchance have need of a fairy telescope to examine them, as we investigate the heavenly bodies.

Illustrations.—1. Having a bowl full of water, it

---

* The word molecule means *a little mass*. A group of atoms forms a molecule, and a collection of molecules constitutes a body. Thus a molecule of water is composed of two atoms of hydrogen and one of oxygen. (See New Chemistry, Rev. Ed., p. 56.)

is easy to add a large quantity of fine salt without apparently increasing the bulk of water in the least. We must only be careful to drop in the salt slowly, giving it time to dissolve **and** the bubbles of air to pass off. When the liquid has taken up all the salt, we can add finely powdered sugar, and afterward other soluble solids in the same manner. In this case we suppose that the particles of sugar are smaller than those of salt, and those in turn smaller than those of water. The **particles of salt** fill the spaces between the particles **of water**, and those of sugar occupy the still smaller spaces left between the particles **of salt**. We may better understand this if we **suppose a bowl** filled with **oranges.** It will still hold **a** quantity of peas, then of gravel, then of fine sand, and lastly some water.

2. At Florence, Italy, in the 17th century, a hollow sphere of gold was filled with water and tightly closed. Pressure was then applied to the outside, and the ball partly flattened. This change of form diminished the size, and so the water was forced through the metal and formed on the surface like drops of dew. This experiment proved that gold has pores, and that they are larger than the molecules of water.

3. In testing large cannon, water is forced into the gun by hydrostatic pressure until it oozes through the thick metal and covers the outside of the gun like froth, then gathers into drops and runs down to the ground in streams.

4. Over the Menai Straits there is a magnificent tubular bridge 100 feet above the water. The tubes were floated to the spot in vessels, and then raised to their position by means of immense Hydraulic Presses. The cylinders of these presses (see R, Fig. 70) were made of iron a foot thick. Yet when in full operation it is said the water would form in drops on the outside, and the workmen, rubbing it off with their fingers, would speak of the machine as "sweating." They were at last compelled to partly stop these pores by mixing oat-meal with the water used in the presses.

5. Stone pillars **and arches** are frequently compressed by the great weight which rests upon them. The columns which support the dome of the Pantheon at Paris are said to have been considerably shortened in this manner.

6. Ashes will "keep fire" because they are por**ous,** and permit enough air to pass in to main**tain** a slow combustion and so preserve the coals **alive.**

7. The process of filtering, so much employed by druggists **and** chemists, depends upon this property; the liquid slowly passes through the pores of the filter, leaving the solid portions behind. Water, in Nature, is thus purified by percolating through beds of sand and gravel. Cisterns for filtering water have a partition in the middle; one side contains charcoal

and sand, the other the rain-water; as the water filters through these substances it is cleansed of its impurities. Small filters are frequently made on the

Fig. 1.

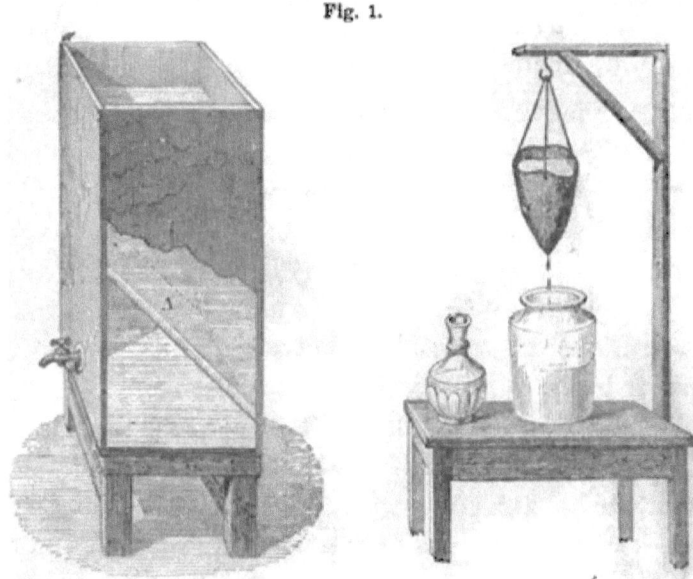

same principle. They consist of a cask nearly filled with gravel and charcoal; the water is poured in a little reservoir at the top and drawn off at the bottom by a faucet.

8. Gases are known to be porous from the fact that when a jar is filled with one kind of gas, it will contain as much of another kind as if the jar were empty. The molecules of the second must spread themselves between the molecules of the first. This illustrates the principle that "one gas is a vacuum for another gas."

Inertia is the property of passiveness. Matter has no power of putting itself in motion when at rest, nor of coming to rest when in motion. A body will never change its place unless moved, and if once started will move forever unless stopped. If we leave our room, and on our return find a book missing, we **know** some one has taken it,—the book could not have gone off at its own suggestion. We generally think a body is more inclined to rest than to motion; and so, while we see how a stone could not throw itself, we find it difficult to understand how, once thrown, it does not stop itself. We shall see hereafter that several forces destroy its motion and bring it to rest.

Illustrations.—1. When we try to start a heavy wagon it requires a great effort, because we have to overcome its inertia, which tends to keep it at rest. When the wagon is in motion it requires as great an exertion to stop it, since then we have again to overcome its inertia, which tends to keep it moving.

2. Inertia causes the danger in jumping from a car when in rapid motion. The body has the speed of the train, while the motion of the feet is stopped by the contact with the ground. One should jump as nearly as he can in the direction in which the train is moving, and with his muscles strained, so as to break into a run the instant his feet touch the ground. Then with all his strength he can gradually overcome the inertia of his body, and after a few rods can turn as he pleases.

Practical Questions.—1. If one is riding rapidly, in which direction will he be thrown when the horse is suddenly stopped? 2. When standing in a boat, why, as it starts, are we thrown backward? 3. When carrying a cup of tea, if we move or stop quickly, why is the liquid liable to spill? 4. Why, when closely pursued, can we escape by dodging? 5. Why is a carriage or sleigh, when sharply turning a corner, liable to tip over? 6. Why, if you place a card on your finger and on top of it a cent, can you snap the card from under the cent without knocking the latter off your finger?

INDESTRUCTIBILITY is the property which renders matter incapable of being destroyed. No particle of matter can be annihilated, except by God, its creator. We may change its form, but we cannot deprive it of existence. Ex.: We cut down a tree, saw it into boards, and build a house. The house burns, and only little heaps of ashes remain behind. Yet in these ashes, and in the smoke of the burning building, exist the identical atoms which have passed through these various forms unchanged in shape, color, or weight.*

Compressibility is often given as a general property of matter. It is, however, a mere result and proof of porosity. It is a distinguishing feature of gases. They are readily compressed; solids require more force, while, for a long time, this property was denied to liquids, and they are even now practically incompressible. Weight is also a property of all

---

* Sir Walter Raleigh, while smoking in the presence of Queen Elizabeth, offered to bet her majesty that he could tell the weight of the smoke that curled upward from his pipe. The bet was accepted. Raleigh quietly finished, and then weighing the ashes, subtracted this amount from the weight of the tobacco he had placed in the pipe; he thus found the exact weight of the smoke. The queen is said to have paid the wager, having in this way learned something of the indestructibility of matter.

bodies with which we are acquainted. It is not, however, essential to our idea of matter, since we can conceive of a substance without weight. Weight is only the result of attraction. Indeed, if there were but one body in the universe it would have no weight, since it could not be attracted in any direction.

THE SPECIFIC PROPERTIES OF MATTER.

THESE are properties which are found only in particular kinds of matter. The most important are Ductility, Malleability, Tenacity, Elasticity, Hardness, and Brittleness. They are doubtless caused by modifications in the attraction of Cohesion, of which we shall soon speak.

DUCTILITY.—A ductile body is one which can be drawn into wire. In the cut is represented a machine

Fig. 2.

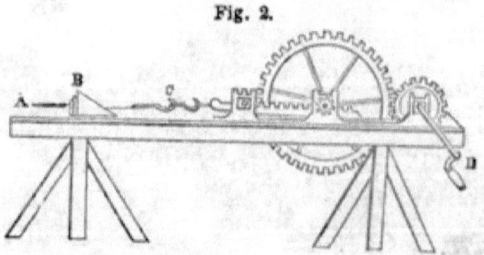

for making iron wire. B is a steel drawing-plate pierced with a series of gradually diminishing holes. A rod of iron, A, is hammered at the end so as to pass through the largest of these. It is then grasped by a pair of pincers, C, and, by turning the crank, D, is drawn through the plate, diminished in size and proportionately increased in length. The speed varies from one to six feet per second, according to size and quality. The rod is then passed through a smaller hole; and the process is continued until the required fineness is reached. The holes in the plate are kept well lubricated with grease or wax. The wire is strongest when drawn cold. After a few drawings the iron loses in part its ductility, and is then annealed by being heated in an oven and afterward cooled slowly. The tenacity of iron is increased by the process of drawing. A bar one inch square, which would sustain 30 tons, on being converted into a coarse wire rope will sustain 40 **tons,** and into fine wire, even 90 tons.

Gold, silver and platinum are the most ductile **metals.** A silver rod an inch thick, covered with gold-leaf, may be drawn to the fineness of a hair and yet retain a perfect coating of gold—3 oz. of the latter metal making 100 miles of the gilt-thread used in embroidery. Platinum **wire has** been drawn so fine that, though **it is the densest of the** metals used for this **purpose,** being nearly three times as heavy as iron, a mile's length weighed only a single grain. (See Revised Chem., page 170, for description of the process.) *Brass wire* is made so small, that **when** woven into gauze there are 67,000 meshes **in a square** inch.

MALLEABILITY.—**A** malleable body is one which can be hammered or **rolled** into sheets. Gold is one of the most malleable of metals. Gold-leaf is prepared in the following manner. An ingot of gold is passed many times between steel rollers, which are so adjusted as to be brought constantly nearer together. An ounce of gold is thus reduced to a ribbon one inch wide and 15 feet long. This is cut into pieces an inch in length. 150 of these are piled up alternately with leaves of strong paper four inches square. A workman with a heavy hammer beats this pile until the gold is spread to the size of the leaves. Each piece is next quartered and the 600 squares are placed between leaves of gold-beater's skin and re-pounded. They are then taken out, spread by the breath, re-cut, and the 2,400 squares re-pounded as before. The beating may be continued until 360,000 leaves make only an inch in thickness. They are finally trimmed and placed between the pages of little books, each of which contains 25 gold leaves.

Copper is so malleable, that it is said that a workman, with his hammer, can beat out a kettle from a solid block of the metal.

TENACITY.—A tenacious body is one which cannot be easily pulled apart. Iron is the most tenacious of the metals. A wire .078 of an inch in diameter, will sustain a weight of nearly 450 lbs., while one of lead would be broken by a weight of 28 lbs.

ELASTICITY is of three kinds: Elasticity of *Compression*, Elasticity of *Expansion*, and Elasticity of *Torsion*, according as a body tends to resume its original form when *compressed*, *extended*, or *twisted*.

Elasticity of Compression.—1. Many *solids* possess

this property in a high degree. A sword was exhibited at the World's Fair in London, which could be bent into a circle, and on being released would fly back and become straight again. The elasticity of ivory may be shown by the following experiment. Spread a thin coat of oil on a smooth marble slab. If an ivory ball be dropped upon it, the size of the impression made will vary with the distance at which the ball is held above the table. This shows that the ivory is flattened, somewhat as is a soap-bubble when it strikes a smooth surface and rebounds. Putty and clay are slightly elastic. 2. *Liquids* are compressed with great difficulty;* but when the force is removed they regain their exact volume. They are therefore perfectly elastic. 3. *Gases* are easily compressed,

Fig. 3.

---

* Thus, for a pressure of one atmosphere, 15 lbs. per square inch, the diminution of volume of the following liquids is only, as compared with the original volume—

1. Water, $\frac{50}{1,000,000}$ 4. Alcohol, $\frac{80}{1,000,000}$
2. Mercury, $\frac{5}{1,000,000}$ 5. Chloroform, $\frac{85}{1,000,000}$
3. Ether, $\frac{15}{1,000,000}$ 6. Sea-water, $\frac{44}{1,000,000}$

but are also perfectly elastic. A pressure of 15 lbs. to the square inch reduces the bulk of water only $\frac{1}{20,000}$ part, whereas it diminishes the volume of a gas one-half. A gas may be kept compressed for years, but will instantly return to its original form on being released.

Elasticity of Expansion.—This property is possessed largely by solids, slightly by liquids, and not at all by gases. Ex.: India-rubber, when stretched, tends to fly back to its original dimensions. When a solid remains stretched for any length of time it loses its elasticity. For this reason a violin is unstrung when not in use. A drop of water hanging to the nozzle of a bottle may be touched by a piece of glass and drawn out to considerable length. When let go, it will resume its spherical form. Gases manifest no tendency to return to their original dimensions when extended.

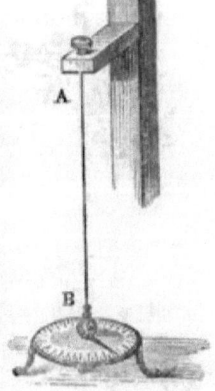

Fig. 4.

Elasticity of Torsion is the tendency of a thread or wire which has been twisted, to untwist again. It is a most delicate test of the strength of a force, and is of great service in accurate measurements in physical science.

HARDNESS.—A hard body is one which does not readily yield to pressure. One body is harder than another when it will scratch or indent it. This property does not de-

A Torsion Balance.

2*

pend on density.* Ex.: Gold is denser than iron, yet is much softer. Mercury is a liquid, yet it is twice as dense as steel.

BRITTLENESS.—A brittle body is one that is easily broken. It is a frequent characteristic of hard bodies. Ex.: Glass will scratch iron, and is extremely brittle.

---

* Density indicates the quantity of matter contained in a given bulk. A dense body has its molecules very closely compacted. The word *rare*, which is the opposite of dense, is generally applied to gases.

Attraction.

"The smallest dust which floats upon the wind
Bears this strong impress of the Eternal mind;
In mystery round it subtle forces roll,
And gravitation binds and guides the whole."

MOLECULAR FORCES.

MOLECULAR FORCES exist in the molecules of matter, and act only at insensible distances.

If we take a piece of iron and attempt to pull it to pieces, we find that there is a force which holds the particles together and resists our efforts. If we try to compress the metal, we find that though there are pores in it and the molecules do not touch each other, yet there is a force which holds the particles apart and resists our efforts as before. If we apply heat, the iron expands and finally melts. If, in like manner, we heat a bit of ice, we notice that the attractive force is gradually overcome, the solid becomes a liquid, and finally the repulsive force predominates and the liquid passes off in vapor. In turn, we can cool the vapor, and convert it back again into water and ice. We thus see that there are two opposing forces which reside in molecules—an Attractive and a Repulsive force, and that the latter is Heat. There are three kinds of the former, *Cohesion*, *Adhesion*, and *Chemical Affinity*.*

---

* Chemical affinity produces chemical changes, and its consideration belongs entirely to chemistry. It binds together atoms or molecules of different kinds, and causes them to form new compounds.

COHESION is the force which holds together molecules of the same kind.

The three states of Matter.—Matter is found in three states, solid, liquid, and gaseous. These depend on the relation of the Attractive and Repulsive forces—Cohesion and Heat. If the attractive force is the stronger, the body is *solid;* if they are nearly balanced, it is *liquid:* if the repulsive force is stronger, it is *gaseous.* Most bodies may be made to take these three states successively. Thus, by the addition of heat, ice may be converted into water and thence into vapor, or *vice versa,* by the subtraction of heat. Most solids pass easily to the liquid form, others go directly from the solid to the gaseous state.

Cohesion acts at insensible distances.—Take two bullets, and having flattened and cleaned one side of each, press them together with a slightly twisting motion. They will be found to cohere when the molecules are crowded into apparent contact. Two panes of plate-glass, which accidentally fall against each other, are thus brought within the range of the attraction of Cohesion, and are frequently cut and polished as one pane. If two globules of mercury be brought near each other, they remain separate until the instant they seem to touch, when they immediately coalesce. Two freshly-cut surfaces of rubber, when slightly warmed and pressed together, cohere as if they formed but one piece.

Welding.—This process illustrates the principle just

named. A rod of iron being broken, we wish to mend it. So we bring the iron to a white-heat at the ends which we intend to unite. This partly overcomes the attraction of Cohesion, and the molecules will move easily upon one another. Laying now the two heated ends upon each other, we pound them with a heavy hammer until over both surfaces the molecules are brought near enough for the attraction of Cohesion to bind them together. Iron and platinum are the only metals which can be welded, as they are the only ones which become softened just before melting. The same property is possessed in a remarkable degree by glass. Gutta-percha, when warmed in water, can also be welded. Dough, wax, and butter can be readily united at common temperatures.

Liquids tend to collect in spheres.—Mix water and alcohol in such proportions that a drop of sweet-oil will fall just to the centre of the fluid. In this way the attraction of the earth is neutralized, and the molecules of the oil are left free to arrange themselves as they please. The drop will form a perfect sphere. The same tendency is seen in dew-drops, rain-drops, globules of quicksilver, in the manufacture of shot, etc.* The reason for this is simply that the force of Cohesion acts toward the centre of the drop. In the spherical body, every portion of the surface is equally distant from the centre; and when that form is assumed every molecule on the

---

* See Fourteen Weeks in Chemistry, Rev. Ed., p. 174.

outside is equally attracted, and there is an equilibrium established.

Solids tend to form regular crystals.—When a liquid becomes a solid, the general tendency is to assume a symmetrical form. The attraction of Cohesion strives to arrange the molecules in an orderly manner. Each kind of matter has its peculiar shape and angle, by which its crystals may be recognized. Even when different substances are contained in the same solution, they separate on crystallization. The beautiful finish and perfection of the crystals thus formed in nature infinitely transcend the workmanship of the highest art. God delights in order as in beauty. Down in the dark recesses of the earth He has fashioned, by the slow processes of His laws, the rarest gems—amethysts, rubies, and diamonds. There are mountain masses transparent as glass, caves hung with stalactites, crevices rich with gold and silver, and lined with quartz. Everywhere we find regularity and symmetry. This tendency is seen in the beautiful crystalline forms of snow-flakes and of frost. A mass of ice seems irregular, yet if closely examined it reveals the perfect crystals crowded together by the rapidity with which the solidification took place. If we watch the surface of water which is slowly freezing, we can see the regular arrangement of the long crystals as they shoot out from each side of the vessel. The very soil is largely composed of broken and decomposed crystals worn down from the rocks by the action of the rain and frost.

Illustrations.—We can illustrate the formation of crystals by adding **alum** to hot water **until** no more will dissolve; then, suspending strings across the dish, setting it away to cool. Beautiful octahedral crystals will collect over the threads and the sides of the vessel. The slower the process the larger will be the crystals. To form the massive crystals found in nature has doubtless required centuries. The large ones seen in the show-windows are made by "feeding" a single small perfect crystal every day with a fresh solution. Melted iron rapidly cooled in a mould has no time to arrange its crystals perfectly. If, however, the iron be **afterward** jarred, as when used for heavy cannon, the axles of rail-cars, etc., the molecules take on the crystalline form and the metal becomes brittle. On examining such a piece of iron we can see in a fresh fracture the smooth, shiny face **of the crystals.**

Tempering and annealing illustrate a curious property of cohesion. A piece of iron is heated and then plunged into oil or water. It becomes hard and brittle. If, on the contrary, it be heated and cooled slowly, it is made tough and flexible. Strangely enough, the same process which hardens iron softens copper. It is supposed that the arrangement of the molecules, and the consequent strength of the metal, depend on the time occupied in cooling. Steel is tempered by heating white-hot, then cooling quickly, **and** afterward re-heating and cooling slowly. The **more it is** re-heated the softer it becomes.

A Prince Rupert's Drop, or Dutch tear, consists simply of a tear of melted glass dropped into water, and so cooled quickly. The outside forms in regular crystals, while the inner portion, not having room to expand, causes a violent strain upon the exterior. The outer shell is strong enough to resist quite a heavy blow with a hammer, but if the small end be nipped off, the whole mass flies into powder with a sharp explosion.

Practical Questions.—1. Why can we not weld a piece of copper to one of iron? 2. Why is a bar of iron stronger than one of wood? 3. Why is a piece of iron, when perfectly welded, stronger than before it was broken? 4. Why do drops of different liquids vary in size? 5. When you drop medicine, why will the last few drops contained in the bottle be of a larger size than the others? 6. Why are the drops larger if you drop them slowly? 7. Why is a **tube stronger than a rod** of the same weight? 8. Why, if you melt scraps of **lead, will they form a** solid mass when cooled? 9. In what liquids is the force **of cohesion greatest?** 10. Name some solids which will volatilize without melting.

ADHESION is the force which holds together molecules of different kinds. Ex.: We fasten together two pieces of wood with glue, two pieces of china with cement, two bricks with mortar, two sheets of paper with mucilage, two pieces of tin with solder, glass and wood with putty, glass and brass with plaster of Paris, and paper to the wall with paste. Paint adheres to the wood-work, **dust** to the wall, and chalk to the blackboard.

The adhesion between animal charcoal and various coloring matters **is** very great. If any liquid containing these substances be filtered through it, the foreign matter in solution will adhere to the charcoal, while the liquid will run through perfectly colorless. Syrup is thus purified by passing through a layer of charcoal 12 or 13 feet thick. The cleansing qualities of common charcoal in water-filters is probably due largely to this property. Bubbles can be blown from soapsuds, because the soap by its adhesive force holds the particles of water together.*

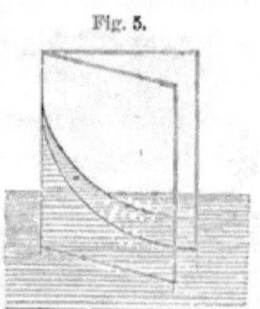

Fig. 5.

Fig. 6.

Capillary Attraction (capillus, a hair) is a variety of adhesion. It may be seen when two plates of glass are placed in water, as shown in Fig. 5, but is exhibited most strikingly in very fine tubes, whence the name.†

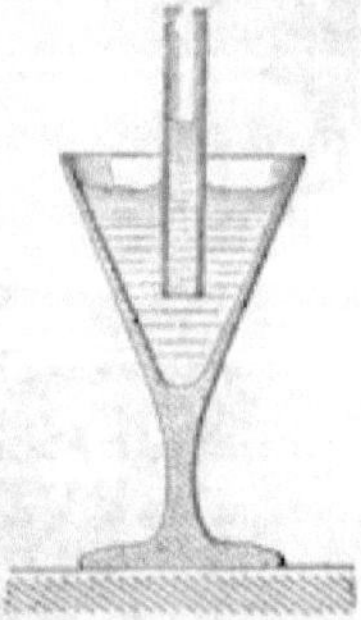

1. If we insert a small glass tube in water, the liquid will rise in the tube. The smaller the tube, the greater will be the height. In this case, it is evident that the adhesive attraction of the glass is greater than the cohesive attraction of the water. There is an *attraction* between the glass and water.

---

* If a bubble be blown at the end of a glass tube, the thin film of water contracting by its cohesive force will frequently drive back the air through the pipe with sufficient strength to extinguish the flame of a candle.

† These tubes may be easily drawn to any length and size, from French-glass tubing, in the heat of a common alcohol-lamp.

2. If we insert a glass tube in a dish of mercury, the capillary action is reversed and the height of the liquid is less than the general level. In this case the adhesive attraction of the glass is less than the cohesive attraction of the mercury. There is an apparent *repulsion* between the glass and mercury.

Fig. 7.

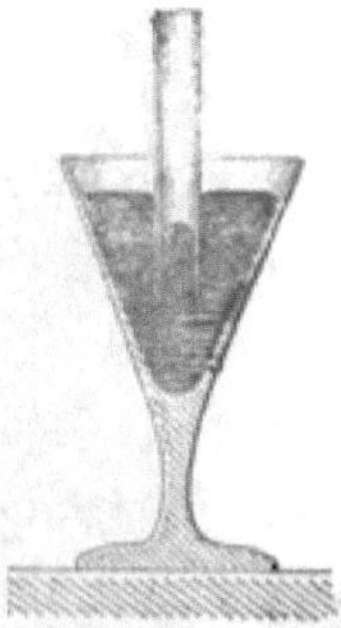

Illustrations.—1. The wick of an oil-lamp or a candle is a bundle of fine capillary tubes or pores which elevate the oil or melted fat and feed the flame. Thus extinguishers are needed to an alcohol-lamp, because by capillary attraction the liquid tends to rise to the top and there evaporate until the lamp is emptied.

2. If the end of a towel be dipped in a basin of water, the whole towel will soon be wet by capillary action through the fine pores and tubes of the cloth. Thus also the capillary tubes of a towel dry one's face after washing.

3. Blotting-paper absorbs ink by means of its capillary tubes.

4. Water poured in the saucer of a flower-pot is elevated through the pores of the earth to the plant.

5. By means of the capillary force water is drawn up through the earth to the surface of the ground, and there moistens the roots of plants and supplies them with the materials of growth. In the winter, when the surface is frozen, the water still finds its way upward,

freezing into ice, which on melting in the spring produces mud, even where there has been but little rain or snow. Ploughing ground causes it to endure drought better, because it stirs the soil and increases the size of the capillary pores, thus partially preventing the water from being carried to the surface and there evaporated.

6. Ropes absorb water by capillary action, swell, and are shortened. Clothes-lines are thus tightened and sometimes broken in a shower. A rope will shrink with such force as to lift a great weight.*

7. Houses are rendered damp by moisture drawn in by the capillary action of the pores in the wood or stone walls.

8. Millstones in Germany are split off by means of wooden wedges. These being driven in when dry, afterward absorb moisture, swell, and burst the

---

* A curious illustration of this is given in the following story. When the great Egyptian obelisk was to be raised in the square of St. Peter's, at Rome, Pope Sixtus V. proclaimed that no one should utter a word aloud until the engineer announced that all danger was passed. As the majestic column ascends, all eyes watch it with wonder and awe. Slowly it rises, inch by inch, foot by foot, until the task is almost completed, when the strain becomes too great. The huge ropes yield and slip. The workmen are dismayed and fly wildly to escape the impending mass of stone. Suddenly a voice breaks the silence. "*Wet the ropes*," rings out clear-toned as a trumpet. The crowd look. There, on a high post, standing on tiptoe, his eyes glittering with the intensity of excitement, is the architect Zapaglia. His voice and appearance startle every one, but his words inspire. He is obeyed. The ropes swell and bite into the stone. The column ascends again, and in a moment more stands securely on its pedestal.

rock, thus saving an immense expenditure of time and money.

Solution.—If we put a little sugar in water, it will dissolve because the adhesive force of the water is stronger than the cohesive force of the sugar. As heat weakens the cohesive force, it commonly hastens solution; and we can dissolve more of a substance, and more rapidly, in hot water than in cold. In like manner pulverizing a solid hastens its solution. A solid will not dissolve in a liquid if there is no adhesion between them. Water absorbs great quantities of the various gases by means of adhesion. It always contains air, which renders it pleasant to the taste. In simply pouring it from one dish to another, we notice that bubbles of air adhering to the stream are carried down, and then rise to the surface and break. It has been proposed to apply this principle to the ventilation of mines. As both pressure and cold weaken the repulsive force of the gases, they favor the adhesion between the molecules of the gases and those of water. Soda-water receives its effervescence and pungent taste from carbonic acid gas, which, being absorbed under great pressure, escapes in little sparkling bubbles as soon as the pressure is removed.

Diffusion of liquids.—Let a tall jar be partly filled with water colored by blue litmus. Then, by means of a long funnel-tube, pour a little clear water containing a few drops of oil of vitriol to the bottom, beneath the colored water. The two will be distinctly defined at first, but in a few days they will mix throughout, as will be seen by the change of color from blue to red. A drop of oil of vitriol may thus be distributed through a quart of water. Most liquids will mingle when brought in contact. If, however, there be no adhesion between their molecules, they will not mix, and will even separate when thoroughly shaken together.

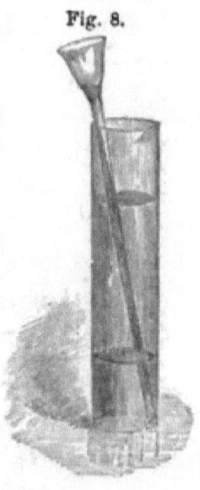

Fig. 8.

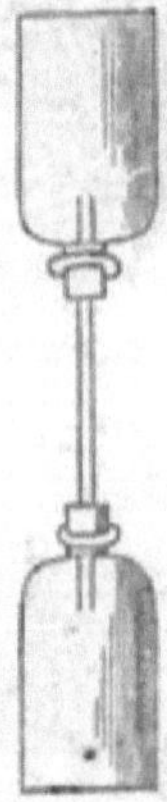

Diffusion of gases.—Hydrogen gas is only $\frac{1}{14}$ as heavy as common air. Yet, if two bottles be arranged as in the figure, the lower one filled with the heavy gas and the upper with the lighter, the gases will soon be found thoroughly mingled.

Fig. 9.

Osmose of liquids.—When liquids are separated from each other by a thin porous substance, they do not mingle uniformly, but the interchange is modified in a most curious manner, according to the nature of the liquid and the substance used. At the end of a glass tube, as in Fig. 10, fasten a bladder full of alcohol. Fill the jar with water, and mark the height to which the alcohol ascends in the tube. The column will soon be found to be gradually rising. On examination we shall see that the alco-

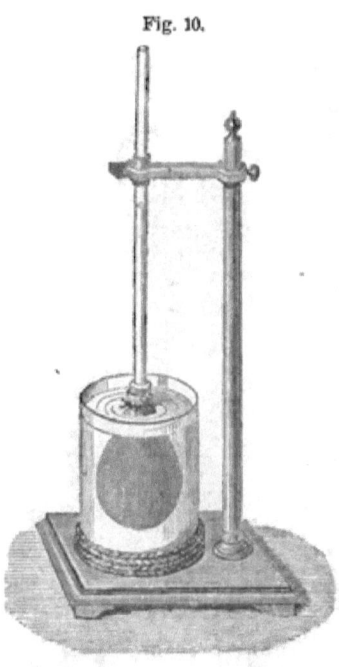

Fig. 10.

hol has been passing out through the pores of the bladder and mixing with the water, while the water has been coming in more rapidly. This has been explained by supposing that the water adheres more strongly than the alcohol to the bladder. Thus, by capillary attraction it is drawn through the membrane, and on the inner side, by the law of diffusion of liquids, mingles with the alcohol. By a similar process some alcohol passes outward and mixes with the water. Whatever liquids are used, that one which wets the membrane most readily will pass through most rapidly. If we should use a collodion balloon, instead of a bladder, the effect would be reversed.

Osmose of gases.—The following experiment would seem to render it probable that there is a similar osmose of gases. Fit a small porous cup, such as is used with Grove's Battery, with a cork and glass tube, as shown in Fig. 11. Fasten the tube so that it will just dip below the surface of the water in the lower jar. Now invert over the porous cup a receiver of hydrogen gas. This gas will pass through the pores of the cup and down the tube so rapidly as to bubble up through the water almost instantly.

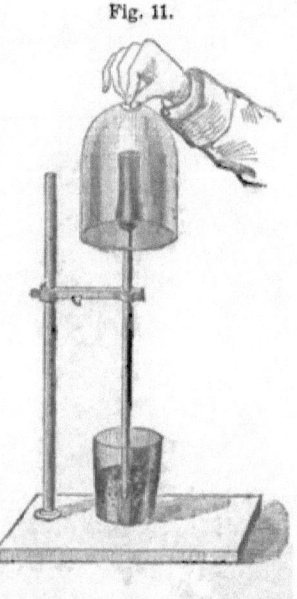

Fig. 11.

Rose balloons, so popular as toys, soon lose their buoyancy, because the hydrogen escapes through the pores of the rubber much more rapidly than the air comes in to take its place. The balloon soon shrinks and drops down from the weight of the rubber.

Practical Questions.—1. Why does cloth shrink when wet? 2. Why do sailors at a boat-race wet the sails? 3. Why does not writing-paper blot? 4. Why does paint prevent wood from shrinking? 5. What is the shape of the surface of a glass of water? One of mercury? 6. Why can we not dry a towel perfectly by wringing? 7. Why will not water run through a fine sieve when the wires have been greased? 8. Why will camphor dissolve in alcohol and not in water? 9. Why will mercury rise in zinc tubes as water will in glass tubes? 10. Why is it so difficult to lift a board out of water? 11. Why will ink spilled on the edge of a book extend farther inside than if spilled on the side of the leaves? 12. If you should happen to spill some ink on the edge of your book, ought you to press the leaves together? 13. Why can you not mix water and oil? 14. What is the object of the spout on a pitcher? *Ans.*—The water would run down the side of the pitcher by the force of Adhesion, but the spout throws it into the hands of Gravitation before Adhesion can catch it. 15. Why will water wet your hand, while mercury will not? 16. Why is a pail or tub liable to fall to pieces if not filled with water or kept in the cellar? 17. Name instances where the attraction of adhesion is stronger than that of cohesion.

ATTRACTION OF GRAVITATION.

WE have spoken of the attraction existing between the molecules of bodies at minute distances. We now notice another form of the same attraction, which acts between masses at all distances.

GRAND LAW OF GRAVITATION.*—Every particle of matter in the universe attracts every other particle of matter with a force directly proportional to its mass, and decreasing as the square of the distance increases.

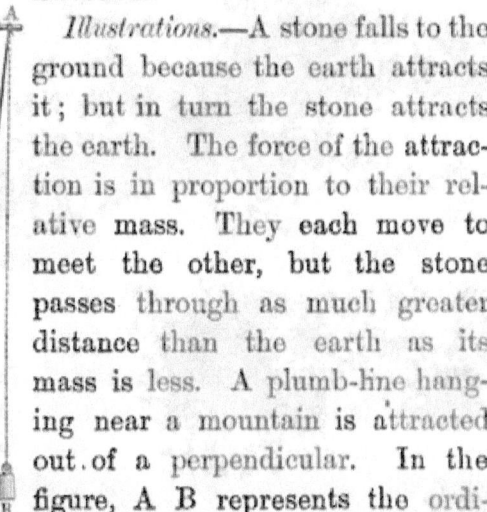

Fig. 12.

Illustrations.—A stone falls to the ground because the earth attracts it; but in turn the stone attracts the earth. The force of the attraction is in proportion to their relative mass. They each move to meet the other, but the stone passes through as much greater distance than the earth as its mass is less. A plumb-line hanging near a mountain is attracted out of a perpendicular. In the figure, A B represents the ordinary position of the line, while A C indicates the attractive power (exaggerated) of the mountain.

---

* See "Fourteen Weeks in Astronomy," p. 34, for history of this law.

This law is not confined to our own world. By it the heavenly bodies are bound to each other, and thus kept in their orbits. It may help us to conceive how the earth is supported, if we imagine the sun letting down a huge cable, and every star in the heavens a tiny thread, to hold our globe in its place, while it in turn sends back a cord to every one. So we are bound to them and they to us. Thus the worlds throughout space are linked together by these cords of mutual attraction, which, interweaving in every direction, make the universe a unit.

GRAVITATION is the general term applied to the attraction that exists between all bodies in the universe. GRAVITY is used to designate the earth's attraction for all terrestrial bodies; it tends to draw them toward the centre of the earth. WEIGHT is the measure of the force of gravity. When we say that a body weighs 10 lbs., we mean that the earth attracts it that amount. The following general principles will explain the various phenomena of weight.

I. *The weight of a body at the centre of the earth* is nothing, because the attraction is there equal in every direction.

II. *The weight of a body above the surface of the earth* decreases as the square of its distance from the centre of the earth increases. Ex.: A body at the surface of the earth (4000 miles from the centre) weighs 100 lbs. What would be its weight 1000 miles above the surface (5000 miles from the centre) of the earth? *Solution*—$(5,000 \text{ m.})^2 : (4,000 \text{ m.})^2 ::$ 100 lbs. : $x = 64$ lbs.

III. *The weight of a body varies on different portions of the surface of the earth.** It will be least at the equator, (1) because, on account of the bulging form of our globe, a body is there pushed out from the mass of the earth, and so removed from the centre of attraction; (2) because the centrifugal force is there the strongest. It will be greatest at the poles, (1) because, on account of the flattening of the earth, a body is there brought nearer its mass and the centre of attraction; (2) because there is no centrifugal force at those points.

FALLING BODIES.—Since the attraction of the earth is toward its centre, all bodies falling freely move in a direct line toward that point. This line is called a *vertical* or *plumb line* (Plumbum, lead, because a lead weight is generally used by mechanics). All plumb-lines point toward the centre of the earth.

Laws of Falling Bodies.—I. *Under the influence of gravity alone, all bodies fall with equal rapidity.*

This is well illustrated by the "Guinea and feather experiment." Let a coin and a feather be placed in a long tube, as in Fig. 13, and the air exhausted. Quickly invert the tube, and the two

---

* In these statements concerning weight, a spring-scale is supposed to be employed. With a pair of balances, the weights used would become heavier or lighter in the same proportion as the body to be weighed.

bodies will fall in the same time. Let in the air again, and now the feather will come fluttering down long after the coin has reached the bottom. Hence we conclude that in a vacuum all bodies descend with equal velocity, and that the resistance of the air is the cause of the variation we see in the falling of light and heavy bodies. The same fact may be noticed in the case of a sheet of paper. When spread out, it merely flutters to the ground; but when rolled together in a compact mass, it falls like lead. In this case we have not increased the force of attraction, but we have decreased the resistance of the air.

Fig. 13.

II. *In the first second, a body gains a velocity of* 32 *feet and falls* 16 *feet.*—This has been proved by experiments with the pendulum, and with Atwood's machine, an instrument constructed with great accuracy for such investigations. It will be noticed that 16 feet, the distance passed through the first second, is the mean between 0, the velocity at the beginning, and 32, the velocity at the close of the second.

III. *In any succeeding second, the velocity is* 16 *feet multiplied by the corresponding even number* 4, 6, 8, *etc.; and the distance is* 16 *feet multiplied by the corresponding odd number,** 3, 5, 7, 9, *etc.*

1. The body commences the second second with a velocity of 32 feet, and as gravity is a constant force, gains 32 feet more during the second, making 64 feet$=4\times 16$ feet. It commences the third second with a velocity of 64 feet, and **gains** 32 feet more; making 96 feet$=6\times 16$ feet. **2.** The mean between 32 feet, the velocity at the beginning of the second second and 64 feet, the velocity at the close, is 48 feet$=3\times 16$ feet. The mean between 64 feet, the velocity at the beginning of the third second and 96 feet, the velocity at the close, is 80 **feet**$=5\times 16$ feet. Hence we conclude that the velocities are as the even numbers, and the distances as the odd numbers.

IV. *In any number of seconds, a body falls* 16 *feet multiplied by the square of the number of seconds.*

We have **just seen that** a body falls 16 feet the first second, and 48 feet the second. Hence in two seconds it falls 16 feet$+48$ feet$=64$ feet$=2^2\times 16$ feet. In three seconds it falls $16+48+80$ feet$=144$ feet$=3^2\times 16$ feet.

Equations of falling bodies.—If we represent the

---

* The odd number corresponding to any second is easily found by doubling the number of the second and subtracting 1 from the result. Ex.: Required the odd number for the eighth second. $8\times 2=16$. $16-1=15$, the eighth odd number.

velocity of a falling body by v, the distance by d, and the time by t, the following equations can be derived from the foregoing laws.

$$\begin{cases} v = 32t & \dots (1). \\ d = 16t^2 & \dots (2). \\ v^2 = 64d & \dots (3). \end{cases}$$

If, now, we let g represent the constant force of gravity, 16 feet in each second, we have from the (3)

$$v = 2\sqrt{gd} \dots (4).$$

Easy way of finding the depth of a well.—Let a stone fall into it, and, with a watch or by the beat of the pulse, count the seconds that elapse before you hear it strike the bottom. Square the number of seconds, and multiply 16 feet by the result. The product is the depth. A little time is required for the sound to come to the ear, but this is so slight that it may be neglected.

When a body is thrown upward the same principles apply, **only** in a reverse manner. Through the influence of **gravity it** loses 32 feet in velocity each second it rises. The velocity necessary in order to elevate it to a certain point, must be that which it would acquire in falling that distance. It will rise just as high in a given time as it would fall in the same time. If a ball be thrown vertically into the air, it will be as long in falling as in rising. In theory, it will strike the earth with the same force with which it was thrown: in practice, however, the ball loses about ⅓ of its force in rising and an equal amount in falling, owing to the resistance of the air.

CENTRE OF GRAVITY.—This is that point on which, if supported, a body will balance itself. The *line of direction* is a vertical line drawn from the centre of gravity. It is the line along which the centre of gravity would pass, if the body should fall. When a body is at rest, the force of gravity which attracts it is said to be in equilibrium.

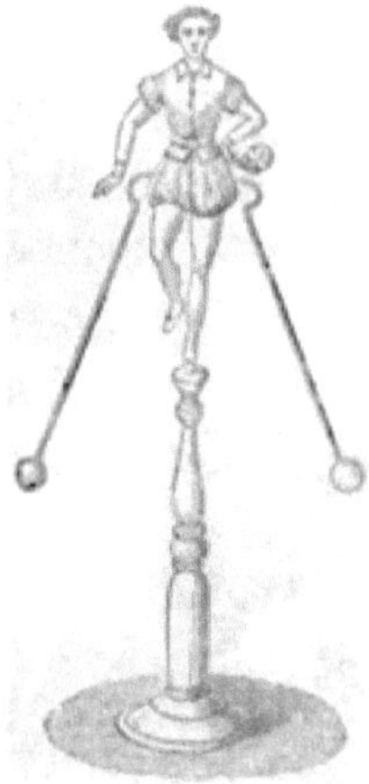

The three states of equilibrium are stable, unstable, and indifferent.

I. A body is in stable equilibrium when the centre of gravity is below the point of support, or when any movement tends to raise the centre of gravity. In Fig. 14 a man has the centre of gravity lowered below the point of support by means of lead balls. Remove these and he immediately falls, but with them he is in stable equilibrium. Any movement tends to raise the centre of gravity, and he returns quickly to a state of rest. The toy in Fig. 15 illustrates a paradox in philosophy,

Fig. 14.

Fig. 15.

viz.: "When a body tends to fall, hang a weight on the heavy side to steady it." A needle may be easily balanced on its point by means of a cork and two jack-knives (Fig. 16), which lower its centre of gravity.

Fig. 16.

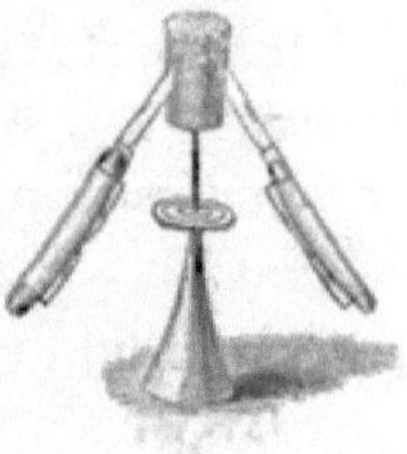

II. A body is said to be in unstable equilibrium, when the centre of gravity is above the point of support, or when any movement tends to lower the centre of gravity. If we take the cork, as balanced in Fig. 16, and invert it, we shall find it very difficult to balance the needle; and, if we succeed, it will readily topple off, because the least motion tends to lower the centre of gravity.

Fig. 17.

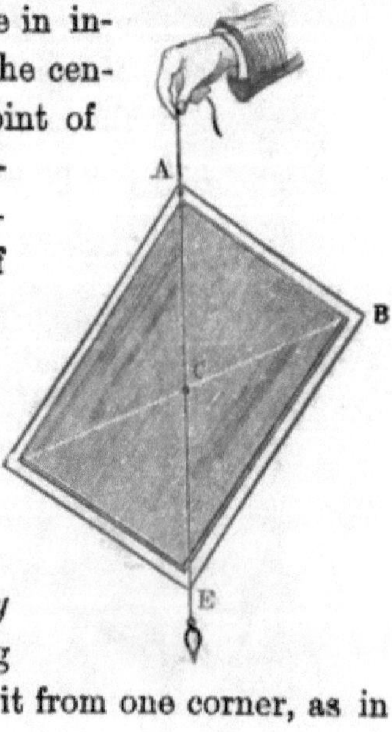

III. A body is said to be in indifferent equilibrium when the centre of gravity is at the point of support, or when any movement tends neither to elevate nor lower the centre of gravity. A ball of uniform density on a level surface will come to rest in any position, because the centre of gravity moves in a line parallel to the floor.

The centre of gravity may be found either by balancing the body or by suspending it from one corner, as in

Fig. 17. By means of a plumb-line, **obtain the line of** direction, A E; then hang it the same way from another corner, and mark the line of direction, B D. The point C, where the two lines cross, is the centre of gravity.

The following general principles will be readily apparent.

a. The centre of gravity in a body always tends to seek the lowest point.

b. A body will never tip over while the line of direction falls within the base, but will do so as soon as it falls without.

c. The higher the centre of gravity must be raised before the line of direction will fall outside of the base, the firmer a body stands.

d. The lower the centre of gravity lies in a body, the more stable it is.

e. In general, narrowness of base combined with height tends to instability; while breadth of base and lowness produce stability. The celebrated leaning tower of Pisa, in Italy, illustrates the principles of gravity. It is about 180 feet high, and its top leans 15 feet, yet the line of direction falls so far within the base that it is perfectly stable, as it has stood for seven centuries. The feeling experienced by a person who for the first time looks down from the lower side of this apparently impending structure, is startling indeed. The towers of Bologna (Fig. 18) are also very wonderful. The lower of these is 130 feet high and is inclined eight feet from the **perpendicular.**

GRAVITATION. 57

Fig. 18.

Leaning Tower at Bologna.

Physiological facts.—Our feet and the space between them form the base on which we stand. By turning our toes outward we increase its breadth. When we stand on one foot, we bend over so as to bring the line of direction within this narrow base. When we carry a pail of water, we balance it by leaning in the opposite direction. When we walk up hill we lean forward, and in going down hill we incline backward, in unconscious obedience to the laws of gravity. We bend forward when we wish to rise from a chair, in order to bring the centre of gravity over our feet; our muscles not having sufficient strength to raise our bodies without this aid. When we walk, we lean forward, so as to bring the centre of gravity as far in front as possible. Thus, walking is a process of falling. When we run, we lean further forward, and so fall faster. (Phys., p. 49.)

THE PENDULUM consists of a weight so suspended as to swing freely. Its movements to and fro are termed *vibrations* or *oscillations*. The path through which it passes is called the *arc*. The extent to which it goes in either direction is styled its *amplitude*. Vibrations performed in equal times are termed *isochronous* (*isos*, equal, and *chronos*, time).

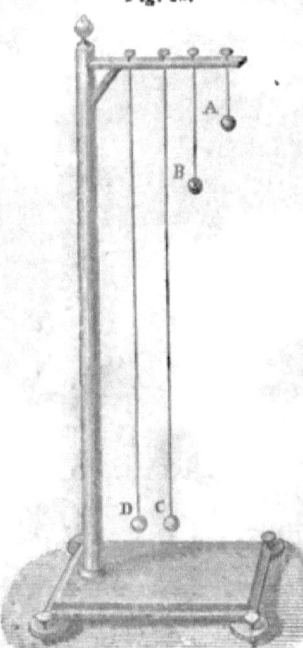

Fig. 19.

1st *Law.*—In the same pendulum, all vibrations of small amplitude are isochronous. If we let one of the balls represented in Fig. 19 swing through a short arc, and count the number of oscillations per minute, we shall find them uniform. This property of the pendulum was discovered by Galileo when a boy, while sitting in the cathedral at Pisa and watching the vibrations of a bronze chandelier which hung from the ceiling. Others had seen this before him. He first noticed that the swinging lamps measured time as well as shed light.

2d Law.—The time of vibration is not affected by the material of which the weight is composed. In Fig. 19, let D be a ball of iron, and C one of wood. They will be found to oscillate together.

3d Law.—The times of the vibrations of different pendulums are proportional to the square roots of their respective lengths. Let A be ¼ the length of C, and it will vibrate three times as fast. If B be ½ the length of C, it will vibrate twice as fast. Conversely, the lengths of different pendulums are proportional to the squares of their times of vibration. A pendulum which vibrates seconds must be four times as long as one which vibrates half-seconds, and sixteen times as long as one which vibrates quarter-seconds. The apparatus represented in Fig. 20, can be used to illustrate very clearly the preceding laws.

Fig. 20.

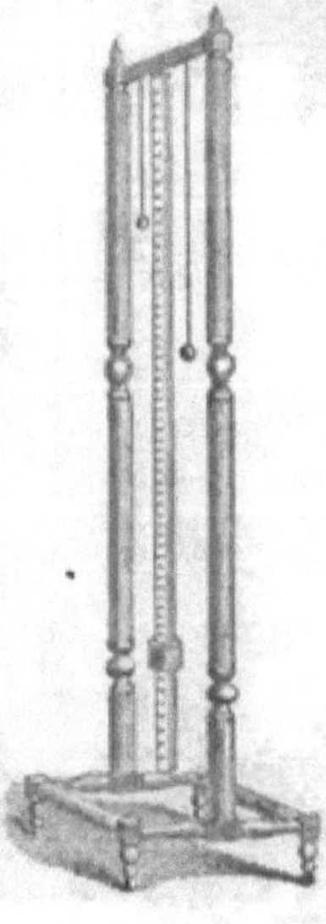

4th Law.—The time of the vibration of the same pendulum will vary at different places on the earth. It will decrease as the square root of the force of gravity increases. At the equator a pendulum vibrates most slowly, and at the poles most rapidly. The length of a second-pendulum at New York is $39\frac{1}{10}$ inches.

Centre of oscillation.—The length of a pendulum is not its absolute length as measured from one extremity to the other, but the distance from the point of support to the centre of oscillation. The upper part tends to move faster than the lower part, and so hastens the speed of the pendulum. The lower part tends to move slower than the upper part, and so retards the speed of the pendulum. Between these two extremes is a point which is neither quickened nor impeded by the rest, but moves in the same time that it would if it were a particle swinging by an imaginary line. This point is called the centre of oscillation. It lies a little below the centre of gravity. In Fig. 21 is shown an apparatus containing pendulums of different shapes, but all having the same absolute length. If they are started together, they will im-

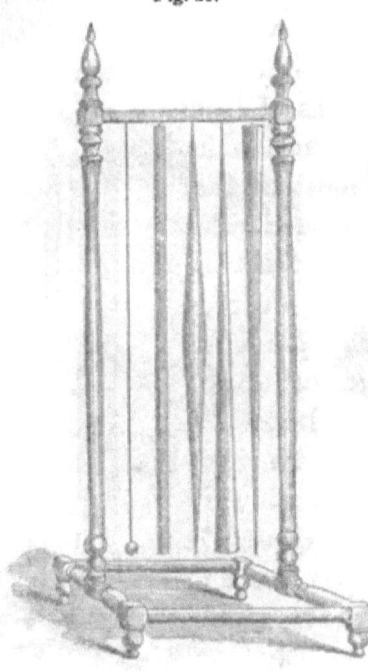

Fig. 21.

mediately diverge, no two vibrating in the same time. As pendulums, they are not of the same length.

The centre of oscillation is found by trial. It has been discovered that the point of suspension and the centre of oscillation are interchangeable. If, therefore, a pendulum be inverted, and a point found at which it will vibrate in the same time as before, this is known to be the centre of oscillation; while the old point of suspension becomes the new centre of oscillation.

The Pendulum as a Time-keeper.—The friction at the point of suspension, and the resistance of the air, soon destroy the motion of the pendulum and bring it to rest. The common clock is simply a machine for keeping up the vibration of the pendulum and counting its beats. In Fig. 22, R is the scape-wheel driven by the force of the clock-weight or spring, and m n the escapement, moved by the forked arm A B, so that only one cog of the wheel can pass at each double vibration of the pendulum. In this manner the

Fig. 22.

oscillations are counted by the cogs on the wheel, while the friction and resistance are overcome by the action of the weight or spring.* As "heat expands and cold contracts," a pendulum increases in length in summer and shortens in winter. A clock, therefore, loses time in summer and gains in winter. To regulate a common clock, we raise or lower the pendulum-bob, L, by means of a nut v at the lower end of the rod.

Fig. 23.

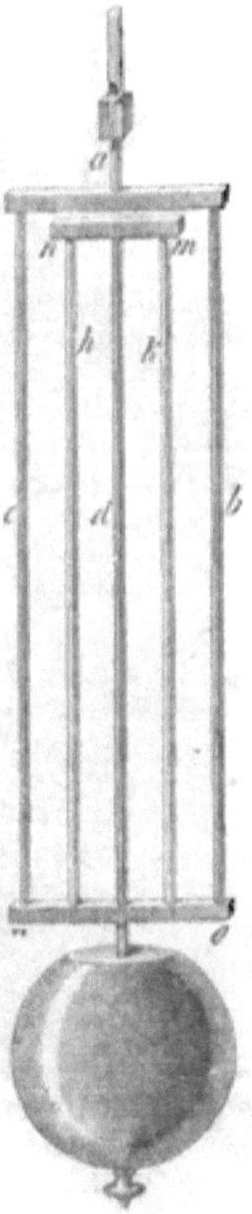

The compensation or gridiron pendulum, consists of several brass and steel rods, which are so connected that the brass, h, k, will lengthen upward and the steel, a, b, c, d, will lengthen downward, and thus the centre of oscillation will be unchanged by any variation in temperature. The mercurial pendulum contains a cup of mercury which expands upward while the pendulum-rod expands downward, and thus keeps the centre of oscillation stationary.

Various uses of the Pendulum.—1. Since the time of the vibration of a pendulum indicates the force of gravity, and since the force of gravity decreases as the square

---

* The action of a clock is best shown by procuring the works of an old clock, and watching the movements of the various parts.

of the distance from the centre of the earth increases, we may, in this manner, find the semi-diameter of the earth at various places, and thus ascertain the figure of our globe. 2. Knowing the force of gravity at any point, the velocity of a falling body can be determined. 3. It may be used as a standard of measures. 4. Foucault devised an ingenious method of showing the revolution of the earth on its axis, founded upon the fact that the pendulum vibrates constantly in *one* plane.

Practical Questions.—**1.** When an apple falls to the ground, how much does the earth rise to meet it? **2.** What causes the sawdust on a mill-pond to collect in large masses? **3.** Will a body weigh more in a valley or on a mountain? **4.** Will a pound weight fall more slowly than a two-pound weight? **5.** How deep is a well if it takes three seconds for a stone to fall to the bottom of it? **6.** Is the centre of gravity always within a body,—as, for example, a ring? **7.** If two bodies, weighing respectively 2 and 4 lbs., be connected by a rod 2 feet long, where is the centre of gravity? **8.** In a ball of equal density throughout, where is the centre of gravity? **9.** Why does a ball roll down hill? **10.** Why is it easier to roll a round body than a square one? **11.** Why is it easier to tip over a load of hay than one of stone? **12.** Why is a pyramid the stablest of structures? **13.** When a hammer is thrown, on which end does it always strike? **14.** Why does a rope-walker carry a heavy balancing-pole? **15.** What would become of a ball if dropped into a hole bored through the centre of the earth? **16.** Would a clock lose or gain time if carried to the top of a mountain? If carried to the North Pole? **17.** In the winter, would you raise or lower the pendulum-bob of your clock? **18.** Why is the pendulum-bob always made flat? **19.** What beats off the time in a watch? **20.** What should be the length of a pendulum to vibrate minutes at the latitude of New York? *Solution*—$(1 \text{ sec.})^2 : (60 \text{ sec.})^2 :: 39.1 \text{ in.} : x = 2.2 + \text{miles}$. **21.** What should be the length of the above to vibrate half-seconds? Quarter-seconds? Hours? **22.** Two pendulums are respectively 16 and 64 inches in length. What is their proportionate time of vibration? **23.** Why, when you are standing erect against a wall, and a piece of money is placed between your feet, can you not stoop forward and pick it up? **24.** If a tower were 198 feet high, with what velocity would a stone, dropped from the summit, strike the ground? **25.** A body falls in 5 seconds; with what velocity does it strike the ground? **26.** How far will a body fall in 10 seconds? With what velocity will it strike the ground? **27.** A body is thrown upward with a velocity of 192 feet the first second; to what height will it rise? (This problem is to be solved as if it read, "How far must a body fall to gain a velocity of 192 feet?") **28.** A ball is shot upward with a velocity of 256 feet; to what height will it rise? How long will it continue to ascend? **29.** Why do not drops of water, falling from the clouds, strike with a force proportional to the laws of falling bodies? *Ans.*—Because they are so small that the resistance of the air nearly destroys their velocity. If it were not for this wise provision, a shower of rain-drops would be as fatal as one of Minié bullets. **30.** Are any two plumb-lines parallel? **31.** A stone let fall from a bridge strikes the water in 3 seconds. What is the height? **32.** A stone falls from a church-steeple in 4 seconds. What is the height of the steeple? **33.** How far would a body fall in the first second at a distance of 12,000 miles above the earth's surface? **34.** A body, at the surface of the earth, weighs 100 tons; what would be its weight 1,000 miles above? **35.** A boy, wishing to find the height of a steeple, lets fly an arrow that just reaches the top and then falls to the ground. It is in the air 6 seconds. Required the height? **36.** A cat let fall from a balloon reaches the ground in 10 seconds. Required the distance? **37.** In what time will a pendulum 40 feet long make a vibration? **38.** Two meteoric bodies in space are 12 miles apart. They weigh respectively 100 and 200 lbs. If they should fall together by force of their mutual attraction, what portion of the distance would be passed over by each body? **39.** If a body weighs 2,000 lbs. upon the surface of the earth, what would it weigh 2,000 miles above? 500 miles above? **40.** At what distance above the earth will a body fall, the first second, $21^{1}/_{2}$ inches? **41.** How far will a body fall in 8 seconds? In the 8th second? In 10 seconds? In the 30th second?

The ancient methods of keeping time were simple indeed. The sun-dial was doubtless the earliest device; afterward the clepsydra was employed. This consisted of a vessel containing water, which slowly escaped into a dish below. In this was a floating body which, by its height, indicated the lapse of time. King Alfred the Great, we read, used candles of a uniform size, six of which lasted a day. He surrounded these with cases of horn as a protection from currents of air. From a mere fancied derivation of this kind, some have spelled the word lantern, lanthorn. Clocks were used in Europe as early as the 11th century. The application of the pendulum was made in the early part of the 17th century. The first clock made in England, about A. D. 1288, was considered of so much importance, that a high official was appointed to take charge of it. The clocks of the Middle Ages were extremely elaborate. They indicated the motions of the heavenly bodies; birds came out and sang songs, cocks crowed and trumpeters blew their horns; chimes of bells were sounded, and processions of dignitaries and military officers, in fantastic dress, marched in front of the dial and gravely announced the time of day. Watches were made at Nuremberg in the 15th century. They were styled Nuremberg eggs. In the 16th century they were in common use. Many were as small as the watches of the present day, while others were as large as a dessert-plate. They had no minute or second hand, and required winding twice per day. They were extremely cumbersome, containing about 800 pieces. In 1658, Dr. Hare invented the main-spring. This gave to watches the accuracy of the pendulum. Waltham watches have but 120 pieces in all. Chronometers are now made so perfect as not to vary a minute in six months.

MOTION.

Motion is a change of place. *Absolute* motion is a change without reference to any other object. *Relative* motion is a change with reference to some other object. *Rest* also is either Absolute or Relative. Ex.: We are in absolute motion with the earth as it flies through space; when we walk, we judge of our motion by the objects around us; a man on a steamer is in motion with regard to the shore, but at rest with reference to the objects on the deck of the vessel. Nothing is in absolute rest. Motion seems to be a law of Nature. *Velocity* is the rate at which a body moves. *Force* is that which **tends to** produce or destroy motion.

Resistances to Motion.—The principal are, Friction, Resistance of the air, and Gravity. (1) Friction is the resistance caused by the surface over which a body moves. It is of two kinds, *sliding* and *rolling*. If the surface of a body could be made perfectly smooth, there would be no friction; but in spite of the most exact polish, the microscope reveals minute projections and cavities. We fill these with oil or grease, and thus diminish friction. Friction, between different bodies, varies curiously. Between like substances it is greater than between unlike. Friction is of great value in common life. Without it, nails, screws, and strings would be useless; **engines could not draw the cars; we could hold nothing in our hands; everywhere we would walk as on glassy ice.** (2) *Resistance of the air.* The resistance which a body meets in passing through air or water is caused by the particles which it must displace. This increases according to the square of the velocity. Thus, if we wish to double our speed in running we must displace twice as much air, and in half the time; hence, the **force** must be quadrupled. (3) *Gravity* **tends to draw all bodies to rest upon the earth.**

MOMENTUM is the quantity of motion in a body. It is equal to the weight of the body multiplied by its velocity per second, expressed in feet. Ex.: A stone, weighing 5 lbs., thrown with a velocity of 20 feet per second, has a momentum of 100 pounds.

The *striking force* of a body is equal to its weight multiplied by the square of its velocity. (See p. 81.) Ex.: A bullet weighing 2 ounces, fired with a velocity of 1,400 feet per second, would strike with a force of 245,000 lbs. Place a hammer on the head of a nail, and, though you push with all your might, you cannot stir it. Swing the hammer by the handle and let it fall upon the nail, and the blow will bury it to the head. On the other hand, a large body may be moving very slowly and yet have an immense momentum. An iceberg, with a scarcely perceptible motion, will crush the strongest man-of-war as if it were an egg-shell. Those who have stood on a wharf have noticed with what prodigious force large vessels grind against each other by the slow movement of the tide. Soldiers have thought to stop a spent cannon-ball by putting a foot against it, but have found its momentum sufficient to break a leg.

Motion is not imparted instantaneously.—We press with all our strength against a large stone. At first it does not stir. But the motion is transmitted from the molecules we touch with our hands, particle by particle, until it reaches the whole body, and the stone yields. A horse will pull at a heavy load for some moments before he starts it; if he should spring forward suddenly, he would be likely to break his harness. It is said that it would require a half minute for a force applied at one end of a mile of railroad-iron to move the last rail. A stone thrown against a pane of glass shatters it; but a bullet fired through it will only make a clean, round hole. The reason is, that the hole is made and the bullet gone before the motion has time to pass into the surrounding particles. A tallow candle may be fired through a board, because it pierces it so quickly that the particles have no time to yield. Its slight cohesion, multiplied by its velocity, is greater than the cohesion of the board.

1st LAW OF MOTION.—*A body once set in motion tends to move forever in a straight line.* This is but another statement of the property of inertia, of which we have already spoken. There is a curious illustration of it seen in the swinging of a pendulum. A pendulum, made to vibrate with the least possible friction, is placed under the receiver of an air-pump. The more perfectly the air is exhausted the longer it will vibrate. In the best vacuum we can produce, it will swing for twenty-four hours. It is supposed that if all the "resistances to motion" were removed, the pendulum would vibrate forever. Philosophers can explain this only on the supposition that a body, **once started,** tends to move forever in a straight **line.** For reasons which are obvious, no experiment can be performed which will directly prove the law. We can see the principle, however, in combination with the second law of motion.

2d LAW OF MOTION.—*A force acting upon a body in motion or at rest, produces the same effect, whether it acts alone or with other forces.*

All bodies upon the earth are in constant motion, and yet we move anything with the same ease that we should, were the earth at rest. We throw a stone directly at an object and hit it, **yet,** within the second, the mark has gone forward many feet.* A ball thrown up into the air with a force that would cause it to rise 50 feet, will ascend to that height whatever horizontal wind may be blowing at the time. If a cannon-ball or a bullet be thrown horizontally, it

---

* The earth moves forward in its orbit about the sun at the rate of 18 miles per second. See Fourteen Weeks in Astronomy, p. 106.

will fall just as fast and strike the earth just as soon as if dropped to the ground from the muzzle of the gun. In Fig. 24, D is an arm driven by a wooden spring, E, and turning on a hinge at C. At D is a hollow containing a bullet so arranged that when the arm is sprung,

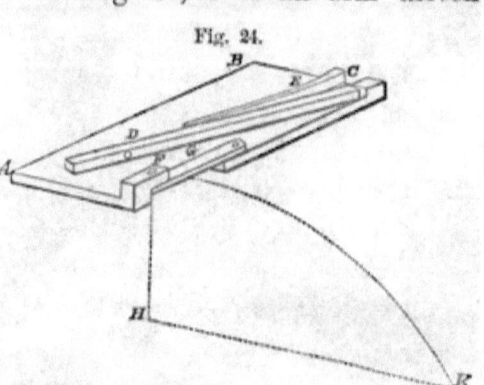

Fig. 24.

it will throw the ball in the line F K. At F is a similar ball, supported by a thin slat, G, and so arranged that the same blow which throws the ball D, will let the ball F fall in the line F H. It will be found that the two balls will strike the floor together. This holds true, no matter how far the ball D may be thrown. We here see that the force of gravity produces the same effect whether it acts alone or in combination with another force.

COMPOUND MOTION.—Let a ball at A be struck by a force which would drive it in the direction A B, and also at the same instant by another which would drive it toward D; the ball will move in the direction A C. The figure A B C D is termed the

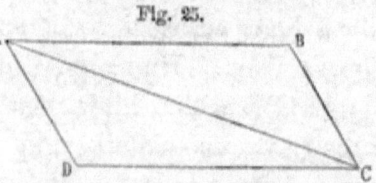

Fig. 25.

"Parallelogram of the Forces," and the diagonal A C the "Resultant."

COMPOSITION OF FORCES.—Whenever a body is acted upon by two forces, we draw lines representing these directions, and mark distances A D and A B, whose lengths represent their comparative velocities. We next complete the parallelogram and draw the diagonal A C, which denotes the resultant of these forces, or the direction in which the body will move. If more than two forces act, we find the resultant of two, then of that resultant and a third force, and so on.

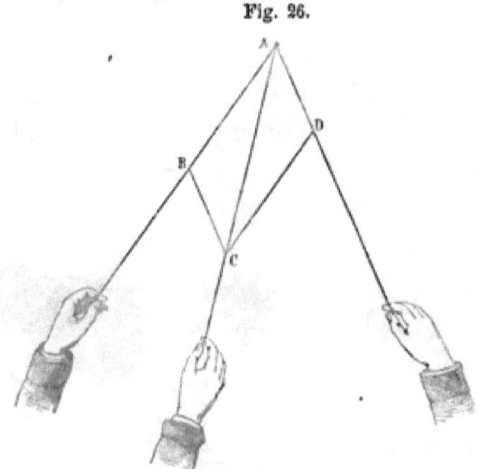

Fig. 26.

Illustrations.—We have many illustrations of compound motion in common life. A person wishes to row a boat across a swift current. It carries him down stream. He steers, therefore, toward a point above that which he wishes to reach, and so goes directly across.—While riding on a car, we throw a stone at some object at rest. The stone, having the motion of the train, strikes just as far ahead of the object as it would have gone had it remained on the train. In order to hit the mark, we should have aimed a little back of it.—The circus-rider wishes, while riding his horse at full speed, to jump through a hoop suspended before him. He simply springs directly upward. He goes forward by the motion which he had when he leaped from the horse. The resultant motion carries him through the hoop and he alights upon the saddle on the other side.—A person riding in a coach drops a cent to the floor. It falls in a vertical line and strikes where it would were the coach at rest.—A bird, beating the air with both its wings, flies in a direction between that of the two.

RESOLUTION OF FORCES. —This is the reverse of the "Composition of Forces." It consists in finding what two forces are equivalent to a given force. It is attained by drawing a parallelogram having the given force for a diagonal. Ex.: There is a wind blowing from the West against G H, the sail of a vessel going North.

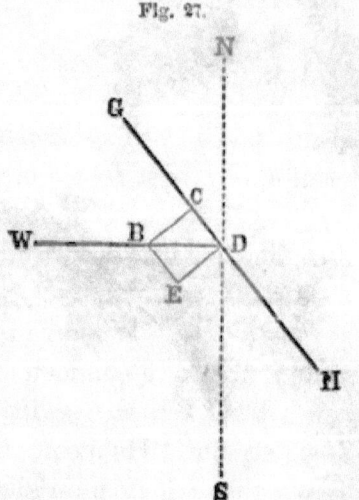

Fig. 27.

We can resolve the wind-force B D into the two forces B E and B C. The former, blowing parallel to the sail, is of no use; the latter is perpendicular to it, and hence tends to drive the vessel before it in a northeast direction. Again, resolving B D in Fig. 28, which represents the vertical force B C in Fig. 27, we find that it is equivalent to two forces, B E and B C. The former pushes the vessel sidewise, but it is mainly counteracted by the shape of the keel and the action of the rudder. The latter is parallel to the course of the ship, and hurries it along northward. By shifting the rigging, one vessel will sail into harbor while another is sailing out, both driven by the same wind. Figs. 29 and 30 show how, by twice resolving the force of the wind from the West, as in the last figures, when the sail G H is placed in the new position, we have (Fig. 30) a force, B C, which drives the vessel southward. If a vessel should wish to sail directly W. against this wind we have supposed, it would *tack* alternately NW. and SW. In this way it could go almost into the "teeth of the wind."

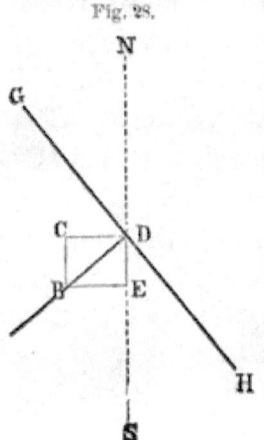

Fig. 28.

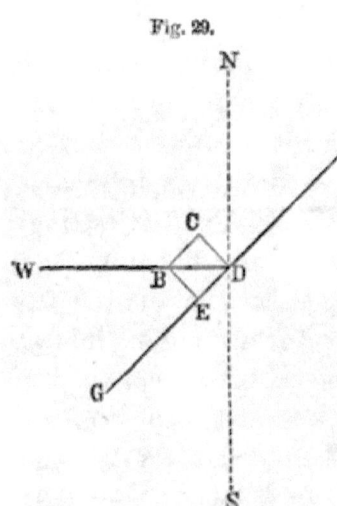

Fig. 29.

In a similar manner we may resolve the three forces which act upon a kite—viz., the pull of the string, the force of the wind, and its own weight. In Fig. 27 let G H represent the face of the kite. We can resolve B D, the force of the wind, into B C and B E. We next resolve B D, in Fig. 28, which corresponds to B C in Fig. 27, into B E and B C. We then have a force, B C, which overcomes the weight of the kite and also tends to lift it upward.

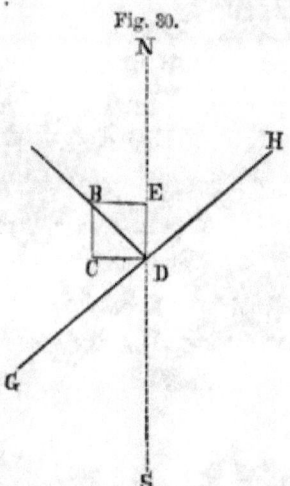

Fig. 30.

The string pulls in the direction D B, perpendicularly to the face. The kite obeys neither one of these forces but both, and so ascends in a direction, D G, between the two. It is really drawn up an inclined plane by the joint force of the wind and the string.

A canal-boat drawn by horses is acted upon by a force which tends to bring it to the bank. This force may be resolved into two, one pulling toward the tow-path, and the other directly ahead. The former is counteracted by the shape of the boat and the action of the rudder; the latter draws the boat forward.

CIRCULAR MOTION is a variety of compound motion produced by two forces called the *Centrifugal* and the *Centripetal*. The former (*centrum*, the centre, and *fugio*, to flee) tends to drive a body from the centre. The latter (*centrum*, the centre, and *peto*, to seek) tends to draw a body toward the centre.

The motion of the heavenly bodies presents the grandest illustration of the operation of these forces. The earth, when first formed, we may suppose, was hurled into space from the hand of the Creator with a force which would send it along the line B C in Fig. 31. According to the law of inertia it would never lose this force, but would continue to move forever in a straight line. Being attracted, however, by the sun in the direction B S, it passes along

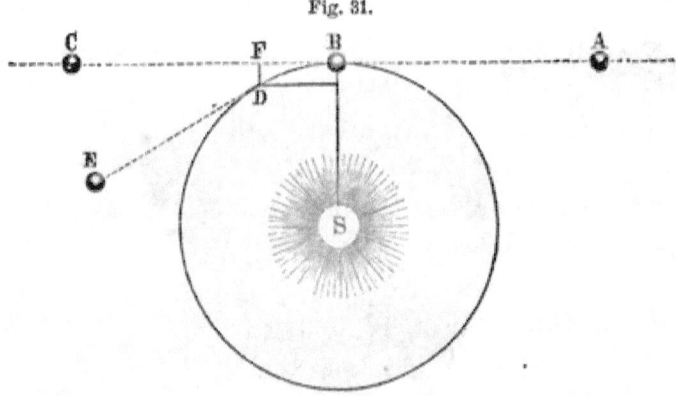

Fig. 31.

the line B D, which is the resultant of these two forces. Should the earth ever lose its own motion, it would fall into the sun and feed that central fire. Should the attraction of the sun cease, it would fly off with headlong speed into the icy, cheerless regions of space.

Examples abound in common life. Water flies from a grindstone on account of the centrifugal force produced in the rapid revolution, which overcomes the force of *adhesion*. In factories, grindstones are sometimes revolved with such velocity that the centrifugal force overcomes the force of *cohesion*, and the ponderous stones fly into fragments. A pail full of water may be whirled around so rapidly that none will spill out, because of the centrifugal force which overcomes the force of *gravity*. When a horse is running around a small circle he bends inward to overcome the centrifugal force.

The rapid revolution of the earth on its axis tends to throw off all bodies headlong into space. As this force acts contrary to that of gravity, it diminishes the weight of all bodies at the equator, where it is greatest, $\frac{1}{287}$. It also tends to drive the water on the earth from the poles toward the equator, and in consequence to heap it up in the equatorial region of the ocean. Were the velocity of the earth's rotation to diminish, the water would run back toward the poles, and tend to restore the earth to a spherical form. This influence is well illustrated by the apparatus shown in the figure. The hoop is made to slide upon its axis, and if it is revolved rapidly it will assume an

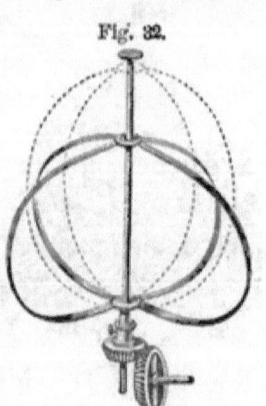

Fig. 32.

oval form, bulging out more and more as the velocity is increased.*

THIRD LAW OF MOTION.—*Action is equal to reaction, and in the contrary direction.* A bird in flying beats the air downward, but the air reacts and supports the bird. The boatman pushes with his pole against the dock, and by the reaction his boat is driven from the shore. The oarsman strikes the water backward with the same force that his boat moves forward. The swimmer kicks with his feet against the water, which reacts and sends him ahead. The powder in a gun explodes with equal force in every direction, driving the gun backward and the ball forward, each with the same momentum. Their relative velocities vary with their respective weights; the heavier the gun the less will the recoil be noticed. When we spring from a boat, unless we are cautious, the reaction will drive it away from the shore. When we jump from the ground, we push the earth from us, while it reacts and pushes us away from it; we separate from each other with equal momenta, and our velocity is as much greater than that of the earth as we are lighter. We cannot jump from a soft surface, because it yields; but a spring-board, which reacts more promptly, aids us. We walk by reason

---

* This apparatus is always accompanied by a variety of objects which may be used to illustrate very beautifully the principle that all bodies tend to revolve about their shortest diameters. This is an assurance that the earth will never change its axis of rotation while it retains its present form.

of the reaction of the ground on which we tread. Thus at every step we take, we cause the earth to move.* The apparatus shown in the figure consists of ivory balls so hung as to readily vibrate. If a ball be let fall from one side it strikes the second ball, which reacts with an equal force and stops the motion of the first, but transmits the motion to the third; that acts in the same manner, and so on through the series, each acting and reacting until the last ball is reached; this reacts and then bounds off, rising as high as the first ball fell (except the loss caused by friction). If two balls be raised, two will fly off at the opposite end; if two be let fall from one side and one from the other, they will respond alternately from either side.

REFLECTED MOTION.—This is produced by the reaction of any surface against which an elastic body is thrown. If a ball be thrown in the direction O B against the surface A C, it will rebound in the line B R. The angle O B P

---

* No force in nature can be wasted. It must accomplish something. "A blow with a hammer moves the earth. A boy could in time draw the largest ship across the harbor in calm weather."

"Water falling day by day
 Wears the hardest rock away."

Statues are worn smooth by the constant kissing of enthusiastic worshippers. Stone steps are hollowed by the friction of many feet. The ocean is filled by small drops which fall from the clouds. We may notice none of these forces singly, but their effects in the aggregate startle us.

(the angle of *incidence*) will be equal to the angle P B R (the angle of *reflection*).

MOTION IN A CURVED LINE.—Whenever two or more instantaneous forces act upon a body, the resultant is a straight line. When one is instantaneous, and the other continuous, it is a curved line. When a body is thrown into the air, unless it be in a vertical line, it is acted upon by the instantaneous force of projection and the continuous force of gravity, and so passes through a line which curves toward the earth.

PERPETUAL MOTION.—Nothing can be more utterly impracticable than to make a machine capable of perpetual motion. No machine can produce power; it can only direct that which is applied to it. In all machinery there is friction; this must ultimately exhaust the power and bring the motion to rest. These principles show the futility of all such attempts.

Practical Questions.—**1.** Can a rifle-ball be fired through a handkerchief suspended loosely from one corner? **2.** A rifle-ball thrown against a board standing edgewise, will knock it down; the same bullet fired at the board, will pass through it without disturbing its position. Why is this? **3.** Why can a boy skate safely over a piece of thin ice, when, if he should pause, it would break under him directly? **4.** Why can a cannon-ball be fired through a door standing ajar, without moving it on its hinges? **5.** Why can we drive on the head of a hammer by simply striking the end of the handle? **6.** Suppose you were on a train of cars moving at the rate of 30 miles per hour; with what force would you be thrown forward if the train were stopped instantly? **7.** In what line does a stone fall from the masthead of a vessel in motion? **8.** If a ball be dropped from a high tower it will strike the ground a little east of a vertical line. Why is this? **9.** It is stated that a suit was once brought by the driver of a light wagon against the owner of a coach for damages caused by a collision. The complaint was that the latter was driving so fast that when the two carriages struck, the driver of the former was thrown forward over the dashboard. On trial he was nonsuited, because his own evidence showed him to be the one who was driving at the unusual speed. Explain. **10.** Suppose a train moving at the rate of 30 miles per hour: on the rear platform is a cannon aimed parallel to the track and in a direction precisely opposite to the motion of the **car**. Let a ball be discharged with the exact speed of the train; where would it fall? **11.** Suppose a steamer in rapid motion, and on its deck a man jumping. Can he jump farther by leaping the way the boat is moving or in the opposite direction? **12.** Why is a "running jump" longer than a standing one? **13.** If a stone be dropped from the masthead of a vessel in motion, will it strike the same spot on the deck that it would if the vessel were at rest? **14.** Could a party play ball on the deck of the Great Eastern when steaming along at the rate of 20 miles per hour, without making allowance for the motion of the ship? **15.** Since action is equal to reaction, why is it not as dangerous to receive the "kick" of a gun as the force of the bullet? **16.** If you were to jump from a carriage in rapid motion, would you leap directly toward the spot on which you wished to alight? **17.** If you wished to shoot a bird in swift flight, would you aim directly at it? **18.** At what parts of the earth is the centrifugal force least? **19.** What causes the mud to fly from the wheels of a carriage in rapid motion? **20.** What proof have we that the earth was once a soft mass? **21.** On a curve in a railroad, one track is always higher than the other. Why is this? **22.** What is the principle of the sling? **23.** The mouth of the Mississippi River is about 2¼ miles farther from the centre of the earth than its source. In this sense it may be said to "run up hill." What causes this apparent opposition to the attraction of gravity? **24.** Is it action or reaction that breaks an egg, when I strike it against the table? **25.** Was the man philosophical who said that it "was not the falling so far, but the stopping so quick, that hurt him?" **26.** If one person runs against another, which receives the greater blow? **27.** Would it vary the effect if the two persons were running in opposite directions? In the same direction? **28.** Why cannot you fire a rifle-ball around a hill? **29.** Why is it that a heavy rifle "kicks" less than a light shot-gun? **30.** A man on the deck of a large vessel draws a small boat toward him. How much does the ship move to meet the boat? **31.** Suppose a string, fastened at one end, will just support a weight of 25 lbs. at the other. Unfasten it, and let two persons pull upon it in opposite directions. How much can each pull without breaking it? **32.** Can a man standing on a platform-scale make himself lighter by lifting up on himself? **33.** Why cannot a man lift himself by pulling up on his boot-straps? **34.** If, from a gun placed vertically, a ball were fired into perfectly still air, where would it fall? **35.** With what momentum would a steamboat weighing 1,000 tons, and moving with a velocity of 10 feet per second, strike against a sunken rock?——(On page 68, we found that a constant force which tends to overcome a continued resistance, like that of air or water, must be as the square of the velocity. This is termed its *living force*, or *vis viva*. This law holds good only in starting bodies from a state of rest and in low velocities. At high rates of speed less force is required. The comparative striking force is, however, always as the square of the velocity.)——**36.** With what momentum would a train of cars weighing 100 tons, and running 10 miles per hour, strike against an obstacle? **37.** What would be the comparative striking force of two hammers, one driven with a velocity of 20 feet per second, and the other 10 feet?

THE ELEMENTS OF MACHINERY.

These are the simple machines to which all machinery can be reduced. The watch, with its complex system of wheel-work, and the engine, with its belts, cranks, and pistons, are only various modifications of some of the six elementary forms—viz., the lever, the wheel and axle, the inclined plane, the screw, the wedge, and the pulley. These six may be still further reduced to the lever and inclined plane. They are termed powers, but do not produce force; they are only methods of applying and directing it. They also enable us to use the forces of Nature, such as wind, water, and steam. The work done by the power is always equal to that done by the weight. The law of all mechanics is—

The power multiplied by the distance through which it moves, is equal to the weight multiplied by the distance through which it moves. Thus 1 lb. of power moving through 10 feet = 10 lbs. of weight moving through one foot.

The Lever is a bar turning on a pivot. The force used is termed the power (P), the object to be lifted the weight (W), the pivot on which the lever turns the fulcrum (F), and the parts of the lever each side of the fulcrum the arms.

The three classes of levers.—I. Power at one end, weight at the other, and fulcrum between. II.

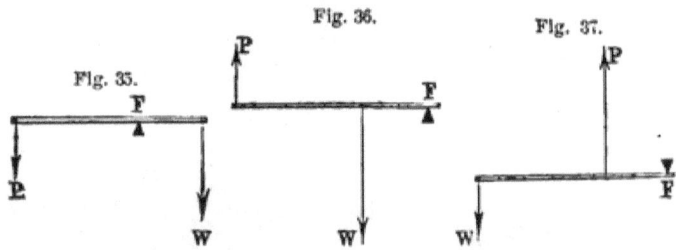

Power at one end, fulcrum at the other, and weight between. III. Weight at one end, fulcrum at the other, and power between.

1st Class of Lever.—We wish to lift a stone. We put one end of a handspike under the stone, and resting the bar on a block at F, we bear down at P. A pump-handle is a lever of the first class. The hand is the P, the water lifted the W, and the pivot the F. A pair of scissors is a double lever of the same class. The cloth to be cut is the W, the hand the P, and the rivet the F.

2d Class.—We may also raise the stone, as in Fig. 39, by resting one end of the lever on the ground, which acts as a fulcrum, and lifting up on the bar. An oar is a lever of the second class. The hand is the P, the boat the W, and the water the F.

3d Class.—The treadle is a lever of the third class. The end, C, resting on the ground is the F, the foot

Fig. 40.

is the P, and the force is transmitted by a rod to the W, the reel above. In the fishing-rod, one hand is the F, the other the P, and the fish the W.

Law of Equilibrium.—The lever is in equilibrium when the arms balance each other. The distance through which the P and the W move depends upon the comparative length of the arms. Let Pd represent power's distance from the F, and Wd weight's distance; then if Pd is twice Wd, the power will move twice as far as the weight. Substituting these terms in the law of Mechanics, we have

$$P \times Pd = W \times Wd, \text{ or } P : W :: Wd : Pd.$$

In the first and second classes, as ordinarily used, we gain power and lose time; in the third class we lose power and gain time.

The Steelyard is a lever of the first class. The

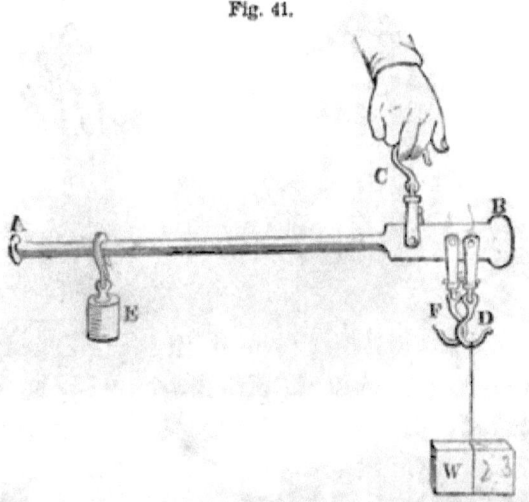

Fig. 41.

power is at E, the fulcrum at C, and the weight at D. If the distance from the pivot of the hook D to the pivot of the hook C is one inch, and from the pivot of the hook C to the notch where E hangs is 12 inches, then a 1-lb. weight at E will balance 12 lbs. at W. If the steelyard be reversed, as in Fig. 42, then the distance of the fulcrum from the W is only ¼ as great, and the same weight at E will balance 48 lbs.

at D. Two sets of notches on opposite sides of the bar correspond to these two positions.

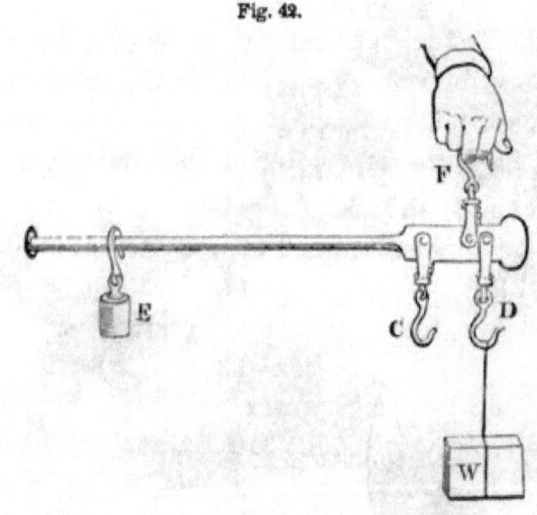

Fig. 42.

The Arm is a lever of the third class. The muscle (Physiology, p. 48) is attached to the bone of the forearm, at a distance of about two inches from the elbow joint, while from the centre of the palm of the hand to the same point is about 13 inches. Hence Wd = 13 inches and Pd = 2 inches. Therefore the force exerted by the muscle must be over six times the weight to be lifted by the hand. What we thus lose in power, we gain in the speed of the motion. We desire to perform quick movements with our hands, and so they are wisely and expressly contrived to meet our wants.

Bent Lever.—In the hammer, when used to draw

a nail, we have a good illustration of a bent lever.

Fig. 43. The real length of the arms is that of the straight lines which correspond to the direction in which the power and weight act with reference to the fulcrum.

The Compound Lever consists of several levers so connected that the short arm of the first acts on the long arm of the second, and so on to the last. If the distance of A from the F be four times that of B, then a power of 5 lbs. at A will lift a W of 20 lbs. at B. If the arms of the second lever are of the same comparative length, then a power of 20 lbs. at C will lift 80 lbs. at E. In the third lever, a power of 80 lbs. at D will, in the same proportion, lift 320 lbs. at G. Thus, with

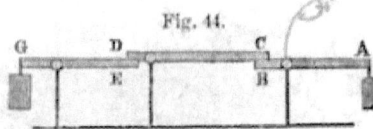

Fig. 44.

this system of three levers, a power of 5 lbs. will balance a weight of 320 lbs. In order, however, to raise the weight one foot, the power must pass through 64 feet. Hay-scales are constructed upon the principle of the compound lever.

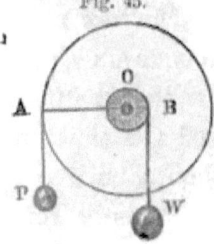

Fig. 45.

THE WHEEL AND AXLE is a modification of the lever. The windlass used for drawing water from a well, is a common form. The power is applied at the handle, the bucket is the W, and the F is the axis of the windlass. The long arm of the lever is the length of the handle, and the short arm is the semi-diameter of the axle. This is seen very clearly in the cross-section shown in Fig. 45, where O is the F, O A the long arm, and O B the short arm. In Fig. 46, instead of turning a handle we take hold of pins inserted in the rim of a wheel. Fig. 47 represents a capstan used on board of vessels for weighing the anchor.

Fig. 46.

The power is applied by means of hand-spikes which radiate outward from the axle. Fig. 48 shows a form of the capstan often used in moving buildings, in which a horse furnishes the power. The wheel and axle has the advantage that it is a kind of per-

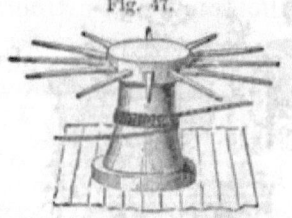

Fig. 47.

petual lever. We are not obliged to prop up the W and readjust the lever, but both arms work continuously.

Law of Equilibrium.—By turning the handle or wheel around once, the rope will be wound once around the axle and the W be lifted that distance. Applying the law of Mechanics, we see that the Power × the circumference of the Wheel = the Weight × circumference of the axle; or, as circles are proportional to their radii,

P : W : : Radius of the Axle : Radius of the Wheel.

If the radius of an axle be 6 inches, and the radius

Fig. 48.

of the wheel 24 inches, then the weight will exceed the power four times.

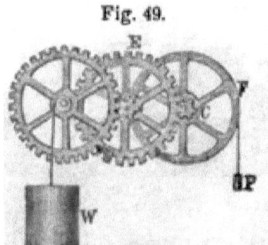

Fig. 49.

Wheelwork consists of a series of wheels and axles which act upon each other on the principle of a compound lever. The cogs on the circumference of the wheel are termed *teeth*, on the axle *leaves*, and the axle itself a *pinion*. If the radius of the wheel F is 12 inches, and that of the pinion 2 inches, then a power of 1 lb. will apply a force of 6 lbs. to the second wheel E. If the radius of this is also 12 inches, then the second wheel will apply a force of 36 lbs. to the third wheel. This, acting on its axle, will balance a W of 216 lbs. In order, however, to lift this amount, according to the principle already named, the weight will only pass through $\frac{1}{216}$ of the distance of the

MECHANICAL POWERS.

93

power. We thus gain power and lose speed. If we wish to reverse this we can apply the power to the axle, and, with a correspondingly heavy power, gain speed. This is the plan adopted in factories, where a heavy water-wheel furnishes abundance of power, and spindles or other swift machinery are to be turned very rapidly.

THE INCLINED PLANE.—If we wish to lift a heavy cask into a wagon, we rest one end of a plank on the wagon-box and the other on the ground. We can then roll the cask up the inclined plane thus

Fig. 50.

formed, when we could not have lifted it directly. When roads are to be made over steep hills, they are sometimes constructed around the hill, like the thread of a screw, or in a winding manner, as shown in Fig. 50. The road from Callao to Lima, in South America, is said to be one of the longest and best-made inclined planes in the world. It is six miles in length, and the total rise is 511 feet. Stairs are inclined planes with steps cut in them to facilitate their ascent.

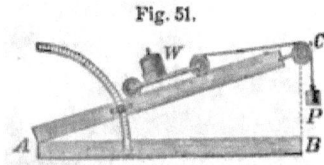

Fig. 51.

Law of Equilibrium.— In Fig. 51 we see that the power must descend a distance equal to A C in order to elevate the weight to the height C B. Applying the law of Mechanics, we have P × length of the inclined plane = W × height of inclined plane; hence,

P : W : : height of inclined plane : length of inclined plane.

Thus, if we roll a barrel of pork, weighing 200 lbs., up a plane 12 feet long and 3 feet high into a wagon, we have x = 50 lbs. : 200 lbs. : : 3 feet : 12 feet. In this case we lift only 50 lbs., or $\frac{1}{4}$ of the barrel, but we lift it through four times the space necessary if we could have raised it directly into the wagon. We thus lose speed and gain power. The longer the inclined plane, the greater the load we can lift, but the longer it will take to do it. If a road ascends one foot in 100 feet, then a horse drawing up a wagon has to lift only $\frac{1}{100}$ of the load, besides overcoming the friction. A body rolling down an inclined plane acquires the same velocity that it would in falling the same height perpendicularly.

A train descending a grade of one foot to 100 feet tends to go down with a force equal to $\frac{1}{100}$ of its weight. Near Lake Lucerne, Switzerland, is a valuable forest of firs on the top of an almost inaccessible Alpine mountain. By means of a wooden trough, the trees are conducted into the water below, a distance of eight miles, in as many minutes. One standing near hears a roar as of distant thunder, and the next instant the descending tree darts past him and plunges downward out of sight. The force with which it falls is so prodigious, that if it jumps out of the trough it is dashed to pieces.

THE SCREW consists of an inclined plane wound around a cylinder. The inclined plane forms the *thread*, and the cylinder the *body*. It works in a *nut* which is fitted with reverse threads to move on the thread of the screw. The nut may turn on the screw, or the screw in the nut. The power may be applied to either as desired, by means of a wrench or a lever. The screw is used in presses for squeezing oil and juices from apples, grapes, rapeseed, linseed, sugar-cane, etc.; for copying letters, for coining money; in vises and in raising buildings.

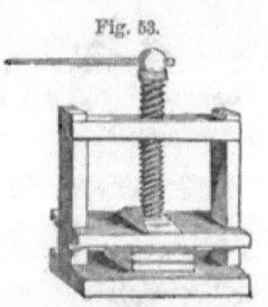

Fig. 53.

Law of Equilibrium.—When the power is applied at the end of a lever, it describes a circle of which the lever is the radius. The distance through which the power passes, is the circumference of this circle; and the height to which the weight is elevated at each revolution of the screw, is the distance between two of the threads. Applying now the law of Mechanics, we have P × circumference of circle = W × interval between the threads; hence,

P : W :: interval : circumference.

The power of the screw may be increased by lengthening the lever, or by diminishing the distance between the threads.

THE WEDGE usually consists of two inclined planes  placed back to back. It is used for splitting logs of wood and blocks of stone; for lifting vessels in the dock; and, in oil-presses, for squeezing. Chimneys which have leaned over, have been righted by wedges driven in on the lower side. Nails, needles, axes, etc., are constructed on the principle of the wedge.

The Law of Equilibrium is, in theory, the same as that of the inclined plane—viz.,

P : W :: thickness of wedge : length of wedge.

In practice, however, this by no means accounts for its prodigious power. Friction, in the other mechanical powers, materially diminishes their efficiency; in this it is essential, since, without it, after each blow the wedge would fly back and the whole effect be lost. Again: in the others, the power is applied as a steady force: in this it is a sudden blow, and is equal to the *momentum* of the hammer.

THE PULLEY is simply another form of the **lever** which turns about a fixed axis or fulcrum. It consists of a wheel, within the grooved edge of which runs a cord.

When we wish to transmit force from one point to another, we may do so either by *pushing* with a rigid bar, or by *pulling* with a flexible cord. The advantage of the latter method is, that we may at the same time change the direction of the force. This is accomplished by a single *fixed pulley*, as in Fig. 55. Here there can be no gain of power or of speed, as the hand F must pull down as much as the weight W, and both move with the same velocity. It is simply a lever of the first class with equal arms. But its use is seen when we remember how, by means of it, a man standing on the ground hoists a flag to the top of a lofty pole, and thus avoids the trouble and danger of climbing up with it. Two fixed pulleys, arranged in the manner shown in Fig. 56, enable us to elevate a heavy load to the upper story of a building by horse-power.

Fig. 55.

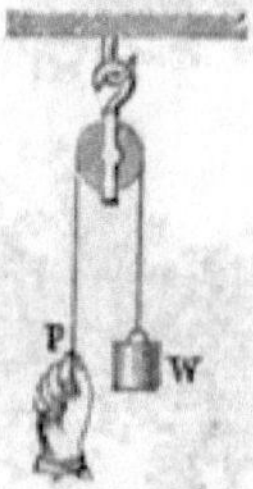

Fig. 56.

Movable Pulley.—A form of the single pulley, where it moves with the W, is represented in Fig. 57. In this, one-half of the barrel is sustained by the hook F, while the hand lifts the other. Since, then, the power is only one-half the weight, it must move through twice the space; in other words, by taking twice the time, we can lift twice as much. Here power is gained and time lost.

We may also explain the action of the single movable pulley by Fig. 58, in which A represents the F, R the W acting in the line O R, and B the P acting in the line B P. This is a lever of the second class; and as B is twice as far from A as O is, the power is only one-half the weight.

Combinations of Pulleys.—1. In Fig. 59, we have the W sustained by three cords, each of which is stretched by a tension equal to the P, hence 1 lb. of power will balance 3 lbs. of weight. 2. In Fig. 60, the power will in the same manner sustain a W of 4 lbs., and must descend 4 inches to raise the W one inch. 3. In the cord marked 1, 1 (Fig. 61), each part has a tension equal to P; and in the cord marked 2, 2,

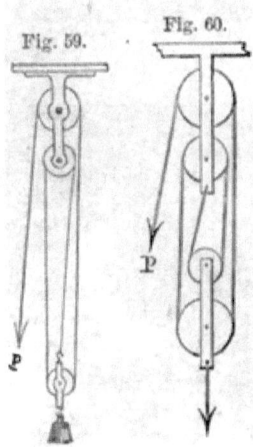

each part has a tension equal to 2 P, and so on with the other cords. The sum of the tensions acting on W is 16, hence W = 16 P.

Fig. 62 represents the ordinary "tackle-block" used by mechanics.

Law of Equilibrium.— In all combinations of pulleys, nearly one-half the effective force is lost by friction.* In most of the forms in use, the W is equal to the P multiplied by twice the number of movable pulleys.

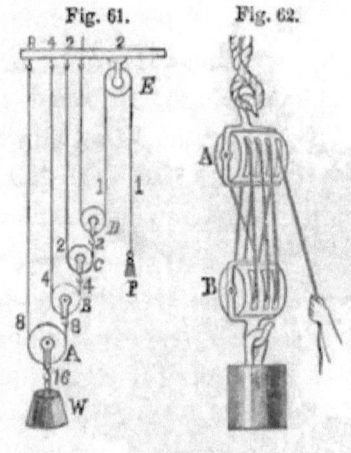

Fig. 61. Fig. 62.

Practical Questions.—**1**. Describe the rudder of a boat as a lever; a door; a door latch; a lemon-squeezer; a pitchfork; a spade; a shovel; a sheep-shears; a poker; a pair of tongs; a balance; a pair of pincers; a wheelbarrow; a man pushing open a gate with his hand near the hinge; a chopping-knife (Fig. 63); a sledge-hammer, when the arm is swung from the shoulder; a nut-cracker. **2**. Show the change that occurs from the second to the third class of lever, when you take hold of a ladder at one end and raise it against a building. **3**. Why is a pinch from the tongs near the hinge more severe than one near the end? **4**. Two persons are carrying a weight of 250 lbs., hanging between them from a pole 10 feet in length. Where should it be suspended so that one will lift only 50 lbs.? **5**. In a lever of the first class, 6 feet long, where should the F be placed so that a power of 1 lb. will balance a W of 23 lbs.? **6**. What power would be required to lift a barrel of pork with a windlass whose axle is one foot in diameter, and handle 3 feet long? **7**. What sized axle, with a wheel 6 feet in diameter, would be required to balance a weight of one ton by a

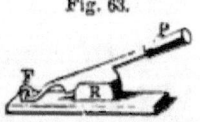

Fig. 63.

---

* The work lost is not destroyed, for this is an impossibility. No force nor matter has been destroyed, so far as we know, since the creation of the world. The force is converted into other forms—heat, electricity, etc., according to the principle of the correlation of forces, p. 316.

power of 100 lbs.* **8.** What number of movable pulleys would be required to lift a W of 200 lbs. by means of a power of 25 lbs.? **9.** How many lbs. could be lifted with a system of 4 movable pulleys and one fixed pulley to change the direction of the force, by a power of 100 lbs.? **10.** What weight could be lifted with a single horse-power* acting on the system of pulleys shown in Fig. 62 (tackle-block)? **11.** What distance should there be between the threads of a screw in order that a P of 25 lbs. acting on a handle three feet long, may lift a ton weight? **12.** How high would a P of 12 lbs., moving 16 feet along an inclined plane, lift a W of 96 lbs.? **13.** I wish to roll a barrel of flour into a wagon, the box of which is four feet from the ground. I can lift but 24 lbs. How long a plank must I get? **14.** The "evener" of a pair of whiffletrees is 3 feet 6 inches in length; how much must the whiffletree be moved to give one horse an advantage of one-third over the other? **15.** In a set of three-horse whiffletrees, having an "evener" 5 feet in length, at what point should the plough-clevis be attached that the single horse may draw the same as each of the span of horses? At what point to give him one-quarter advantage? **16.** What weight can be lifted with a power of 100 lbs. acting on a screw having threads one-quarter of an inch apart, and a lever handle 4 feet long? **17.** What is the object of the big balls always cast on the ends of the handle of the screw used in presses for copying letters? **18.** In a pair of steelyards 2 feet long, the distance from the weight-hook to the fulcrum-hook is 2 inches; how heavy a body can be weighed with a 1-lb. weight at the further end? **19.** Describe the change from the first to the third class of levers, in the different ways of using a pitchfork or spade. **20.** Why are not blacksmiths' tongs and fire-tongs constructed on the same principle? **21.** In a lever of the third class, what W will a P of 50 lbs. balance, if one arm is 12 feet and the other 8 feet long? **22.** In a lever of the second class, what W will a P of 50 lbs. balance, with a lever 12 feet long, and W 3 feet from the F? **23.** In a lever of the first class, what W will a P of 50 lbs. balance, with a lever 12 feet long, and the F 3 feet from the W? **24.** In a wheel and axle, the P = 40 lbs., the W = 360 lbs., and the diameter of the axle = 8 in. Required the circumference of the wheel. **25.** Suppose, in a wheel and axle, the P = 20 lbs., the W = 240 lbs., and the diameter of the wheel = 4 feet. Required the circumference of the axle? **26.** Required, in a wheel and axle, the diameter of the wheel, the diameter of the axle being 10 inches, the P 100 lbs., and the W 1 ton? **27.** What P would be necessary to sustain a W of 3,780 lbs. with a system of six movable pulleys, and a single rope passing over them all? **28.** How many movable pulleys would be required to sustain a W of 420 lbs., with a P of 210 lbs.?

---

* A horse-power is reckoned in Mechanics as a force which will lift 32,000 lbs. one foot in one minute, without any assistance of machinery.

The Pressure of Liquids and Gases.

"The waves that moan along the shore,
 The winds that sigh in blowing,
Are sent to teach a mystic lore,
 Which men are wise in knowing."

HYDROSTATICS.

HYDROSTATICS treats of liquids at rest. Its principles apply to all liquids; but water, on account of its abundance, is taken as the type of the class, and all experiments are based upon it.

I. LIQUIDS TRANSMIT PRESSURE EQUALLY IN ALL DIRECTIONS.—This is the first and most important law. As the particles of a liquid move freely among themselves, there is no loss by friction, and any force will be transmitted equally, upward, downward, and sidewise. Thus if a bottle be filled with water and a pressure of 1 lb. be applied upon the cork, it will be communicated from particle to particle throughout the water. If the area of the cork be one square inch, the pressure upon any square inch of

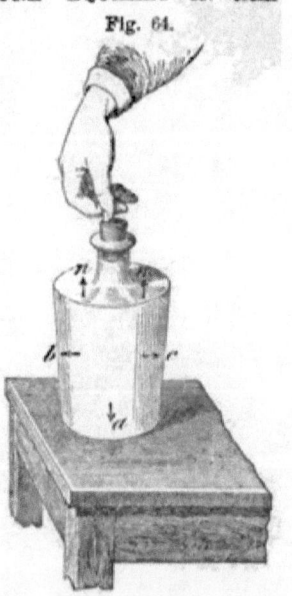

Fig. 64.

the glass at *n*, *a*, *b* or *c*, will be equal to 1 lb. If the inside surface of the bottle be 100 square inches, then a pressure of 1 lb. upon the cork will produce a total force of 100 lbs., tending to burst the bottle.

Illustrations of the transmission of pressure by liquids. — Under some circumstances this is more perfect than that by solids. Let a straight tube, A B, be filled with a cylinder of lead, and a piston, be fitted to the end of the tube. If now a force be applied at O it will be transmitted without loss to P. If, instead, we use a bent tube, the force will be transmitted in the line of the arrow, and will act upon P but slightly, if at all. If, however, we fill the tube with water, the force will pass without diminution. With cords, pulleys, levers, etc., we always lose about one-half of the force by friction; but this "liquid rope" transmits it with

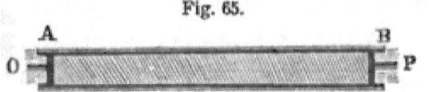
Fig. 65.

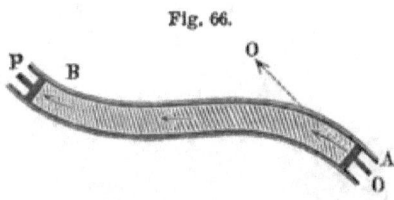
Fig. 66.

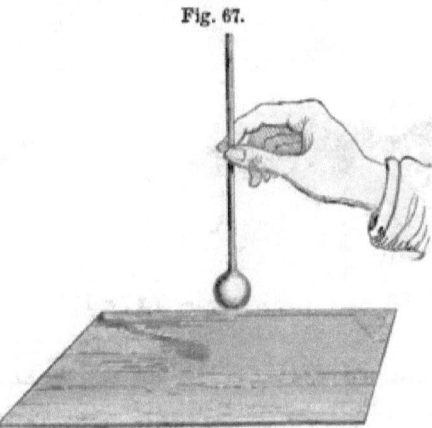
Fig. 67.

no sensible loss. Take a glass bulb and stem, as shown in Fig. 67, and fill it with water by the process explained under Thermometers. When full, if you are careful to let the stem slip loosely through the fingers as the bulb strikes, you may pound with it upon a smooth surface with all your strength. In this case, the force of the blow is instantly transmitted from the thin glass to the water, and that being almost incompressible, makes the bulb nearly as solid as a ball of iron.

If a Rupert's drop be held in a vial of water, as in Fig. 68, and the tapering end be broken, the force of the concussion will be transmitted to all parts of the glass and the vial will be instantly shattered.

Fig. 68.

Water as a mechanical power.—Take two cylinders, P and *p*, connected as in Fig. 69, fitted with pistons and filled with water. Let the area of *p* be 2 inches and that of P be 100 inches. Then, according to the principle of the equal pressure of liquids, a downward pressure of 1 lb. on each square

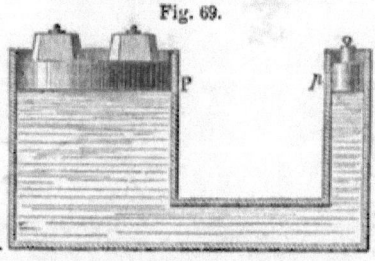

Fig. 69.

inch of the small piston will produce an upward pressure of 1 lb. on each square inch of the large piston. Hence a power of 2 lbs. would lift a weight of 100 lbs. This proportion may be increased by diminishing the size of p and increasing that of P,

Fig. 70.

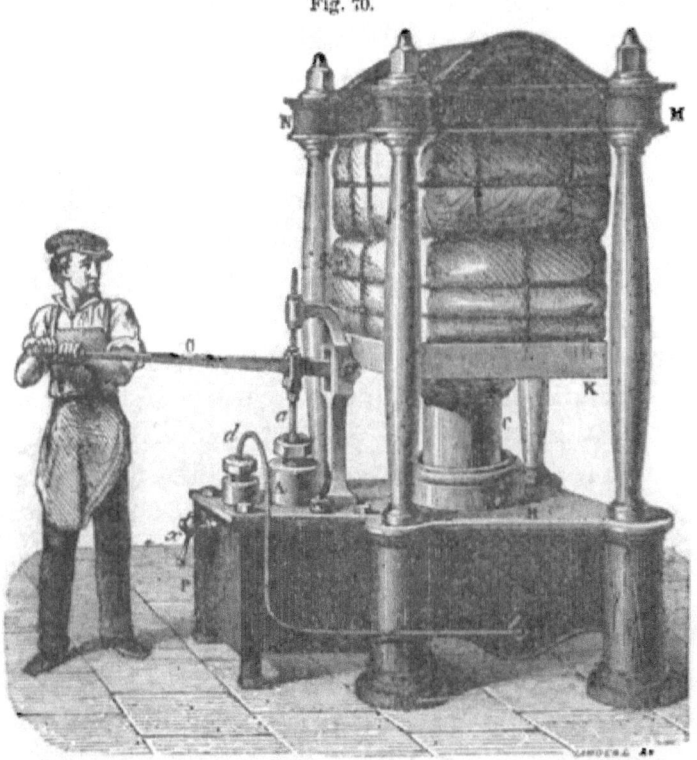

so that the weight of a girl's hand could lift a man-of-war. Water has been well termed the "seventh mechanical power."

Hydrostatic Press.—Fig. 70 represents a press constructed on the principle just explained. As the piston a is forced down upon the water in the cylinder A by the workman, the pressure is transmitted through the bent tube of water d around under the large piston C which lifts up the platform K, and thus compresses the bales placed upon it. If the area of a is 1 inch and that of C 100 inches, then a force of 100 lbs. will lift 10,000 lbs. Still further to increase the efficiency of this press, the handle is a lever of the second class. If the distance of the hand from the pivot is ten times that of the piston, a P of 100 lbs. will produce a force of 1,000 lbs. at a. This will become 100,000 lbs. at C. Hence, with a press of this size, a power of 100 lbs. will lift a weight or produce a pressure of 100,000 lbs. Applying the principle of Mechanics, we see that here as elsewhere there is no force created, but that P $\times$ Pd = W $\times$ Wd. The platform will ascend only $\frac{1}{100,000}$ part of the distance the hand descends. This machine is used for baling hay and cotton for transportation; for launching vessels; for testing the strength of ropes, chains, etc. The presses employed for raising the immense tubes of the Britannia Bridge were each capable of lifting 2,672 tons, or of throwing water in a vacuum to a height of nearly six miles.

II. Liquids influenced by gravity alone.—In this case there is no external pressure applied. The lower parts of a vessel of water must bear the weight of the upper parts. Thus each particle of water at rest is pressed downward by the weight of the minute column it sustains. It must, in turn, press in every direction with the same force, else it would be driven out of its place and the liquid would no longer be at rest. Indeed, when a liquid is disturbed in any manner it comes to rest; *i. e.*, there is an equilibrium established only when there is this equality of pressure produced. In consequence of this constant pressure the following laws obtain:

1st. *Liquids at rest press downward, upward, and sidewise with the same force.*—This may be illustrated by the following experiment. If the series of glass

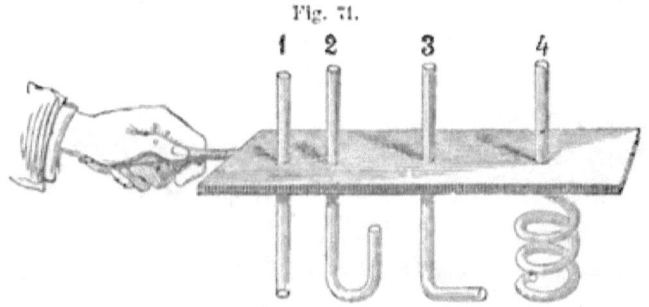

Fig. 71.

tubes shown in Fig. 71 be placed in a pail of water, the liquid will be forced up 1 by the upward pressure of the water, 2 by the downward pressure, 3 by the lateral pressure, and 4 by the three combined in different portions of the tube. The water will rise in them all to the same height—*i. e.*, to the level of the water in the pail.

2d. *The pressure increases with the depth.*—The pressure at the depth of one foot is the weight of one cubic foot of water—viz., 62½ lbs. (1,000 oz.); at

HYDROSTATICS.

2 feet, twice that amount; and so on.* In sea-water it is greater, as that weighs 64.37 lbs. per cubic foot. At great depths this pressure becomes enormous. If a strong square glass bottle, empty and firmly corked, be sunk into the water, it will generally be crushed inward before it sinks ten fathoms. It is said that the Greenland whale sometimes descends to the depth of a mile, but always comes up exhausted and blowing blood. When a ship founders at sea, the great pressure forces the water into the pores of the wood, so that no part can ever rise again to the surface to reveal the fate of the lost vessel.

3d. *The pressure does not depend on the shape or size of the vessel.*—In the apparatus shown in Fig. 72 the water rises to the same height in the variously shaped tubes, which communicate with each other, whatever may be their form or size. If more water be poured in one, it will rise higher in all the others.

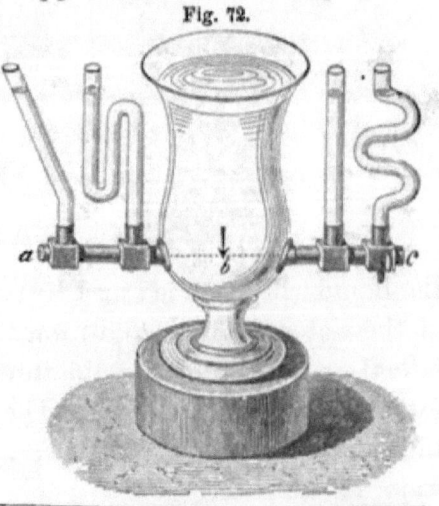

Fig. 72.

* Depth.	Lbs. per sq. foot.	Depth.	Lbs. per sq. foot.
1 ft.	62.5	100 ft.	6,250
10 ft.	625.	1 mile,	330,000
16 ft.	1,000.	5 miles,	1,650,000

The Hydrostatic Bellows consists of two boards, each hinged on one side and resting on a rubber bag, to which is attached an upright tube, A.—Water is poured in at A until the bag and tube are filled. The pressure of the column of water in the tube lifts the weights hung by crossbars beneath. Whether we use the tube A or B will make no difference in the weight supported, although the former holds ten times as much water as the latter. The tube C, however, being much longer, will exert a greater pressure.

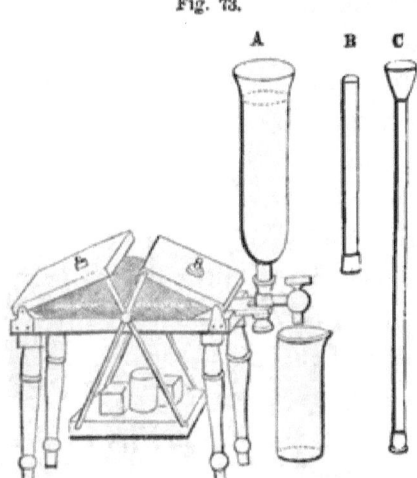

Fig. 73.

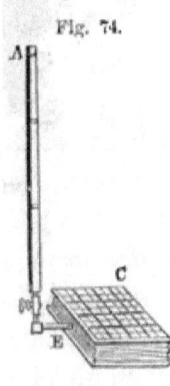

Fig. 74.

Another form of the same apparatus (Fig. 74) consists of two boards connected by a band of leather, in which a tall tube A is inserted. If this be filled with water, the pressure will be sufficient to lift a weight as much greater than the weight of the water in the tube as the area of the bellows-board is greater than the area of the tube. Applying again the principle of Mechanics, we see that if one ounce of water should raise a weight of 50 oz. one inch, then the water must fall 50 inches.

Fig. 75.

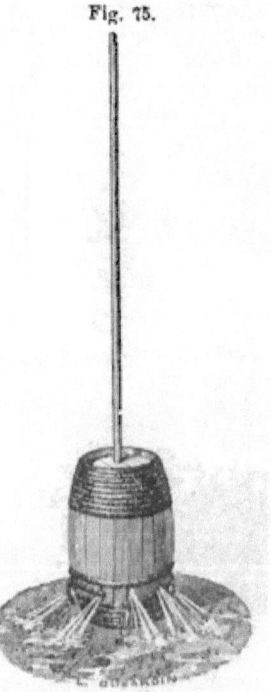

A strong cask fitted with a small pipe 30 or 40 feet long, if filled with water will burst asunder. The pressure is as great as if the tube were of the same diameter as the cask. In a coffee or tea pot the small quantity of liquid in the spout balances the large quantity in the vessel. If it were not so, it would rise in the spout and run out.

The principle that a small quantity of water will thus balance another quantity, however large, or will lift any weight, however great, is frequently termed the "Hydrostatic Paradox." We see, however, that it is only an instance of the general law.

4th. *Water seeks its level.*—This tendency is seen especially in fountains and in the supply of water furnished to cities from an elevated reservoir. In Fig. 76 the tank is situated on a hill at the left, whence the water is conducted underground through a pipe to the fountain. The jet will rise, in *theory*, to the level of the surface, but in *practice* it falls short of this, owing to the friction at the nozzle of the pipe and in passing through the air, and the weight of the falling drops. It has been thought that the Romans knew nothing of this property of liquids, because they built immense stone aqueducts a hundred miles in length, spanning valleys and rivers at vast

Fig. 76.

expense. Modern engineers simply carry the water in pipes through the valley or under the bed of the river, knowing that it will rise on the opposite side to its level. The ancients appear to have understood this principle, but could not make pipes capable of resisting the pressure.

Artesian wells are so named because they have been used for a long time in the province of Artois, in France. They were, however, employed by the Chinese from early ages for the purpose of procuring gas and salt water.

Let A B and C D represent curved strata of clay impervious to water, and K K a layer of gravel and fine sand. The rain falling on the distant hills filters down to C D, and collects in this hollow

Fig. 77.

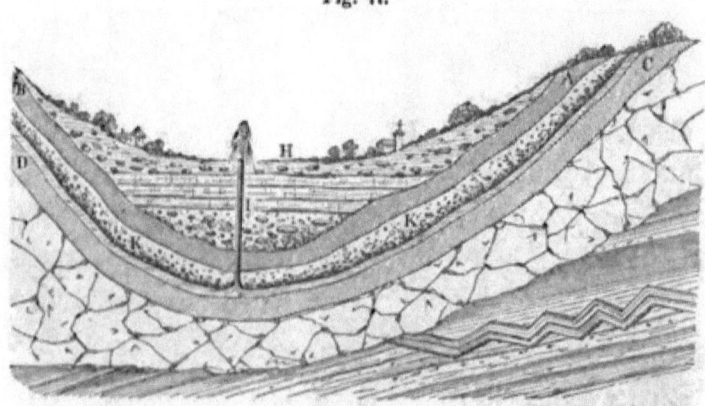

basin. If a well be bored at H, as soon as it reaches the stratum of gravel beneath, the water will rush upward, under the tremendous lateral pressure, to the height of the source, and often spout high in the air. The well at Grenelle, near Paris, is very celebrated. It is at the bottom of a great chalk-basin which extends many miles from the city. It is over 1,800 feet deep and furnishes 1,000,000 gallons daily. The wells of Chicago, on the level prairie, are about 700 feet deep, and discharge daily about 1,250,000 gallons of clear cold water. The force with which the water comes to the surface indicates a head of 125 feet above Lake Michigan. Its source must be far away beyond Lake Superior, perhaps even beyond the Mississippi, toward the Rocky Mountains. Artesian wells are bored in the sands of Sahara; gardens are planted and dates flourish wherever water is supplied. Brigades of engineers are thus pushing forward the conquest of the African desert.

RULES FOR THE CALCULATION OF PRESSURE.—1. *To find the pressure on the bottom of a vessel.* Multiply the area of the base by the perpendicular height, and that product by the weight of a cubic foot of the liquid.—2. *To find the pressure on the side of a vessel.* Multiply the area of the side by half of the perpendicular height, and that product by the weight of a cubic foot of the liquid.

The pressure on the bottom of a cubical vessel full of water, is the weight of the water: on each

Fig. 78.

side, one-half; and on the four sides, twice the weight; therefore on the five sides, the pressure is three times the weight of the water.

THE WATER-LEVEL.—The surface of standing water is said to be level—*i. e.*, horizontal to a plumb-line. This is true for small sheets of water, but for larger bodies an allowance must be made for the circular figure of the earth. The curvature is 8 inches per mile; $2^2 \times 8$ inches = 32 inches for two miles; $3^2 \times 8$ inches = 72 inches for three miles, etc. The *spirit-level* is an instrument used by builders for

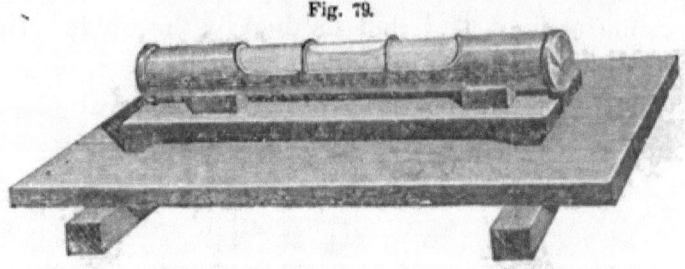

Fig. 79.

levelling. It consists of a slightly curved glass tube so nearly full of alcohol that it holds only a bubble of air. When the level is horizontal, the bubble remains at the centre of the tube.

SPECIFIC GRAVITY is the weight of a substance compared with the weight of the same bulk of another substance. It is really a method of finding the density of a body. Water is taken as the standard[*]

---

[*] A cubic inch of distilled water at a temperature of 62° F., with the barometer at 30 inches. This standard weighs 252.456 grs.: 7,000 grs. make a pound **Avoirdupois** and 58,333 a gallon.

for solids and liquids, and air for gases. A cubic inch of zinc weighs seven times as much as a cubic inch of water; hence its specific gravity = 7. A cubic inch of carbonic acid gas weighs 1.52 times as much as the same volume of air; hence its specific gravity = 1.52.

Buoyant Force of Liquids.—The cube $a\ b\ c\ d$ is immersed in water. We see that the lateral pressure at a is equal to that at b, because both sides are at the same depth; hence the body has no tendency toward either side of the jar. The upward pressure at c is greater than the downward pressure at d, because its depth is greater; hence the cube has a tendency to rise.

Fig. 80.

This upward pressure is called the buoyant force of the water. Its law, discovered by Archimedes, is—

The buoyant force is equal to the weight of the liquid displaced. The downward pressure at d is the weight of a column of water whose area is that of the top of the cube, and whose perpendicular height is $n\ d$: the upward pressure at c is equal to the weight of a column of the same size whose perpendicular height is $c\ n$. The difference between the two, or the buoyant force, is the weight of a bulk of water equal to the size of the cube.

The same is shown in what is called the "cylinder

and bucket experiment." The cylinder a exactly fits in the bucket b. The glass vessel in which the bucket hangs is empty. The apparatus is balanced by weights placed in the scale-pan. Water is then poured into the glass vessel. Its buoyant force will raise the cylinder and depress the opposite scale-pan.

Fig. 81.

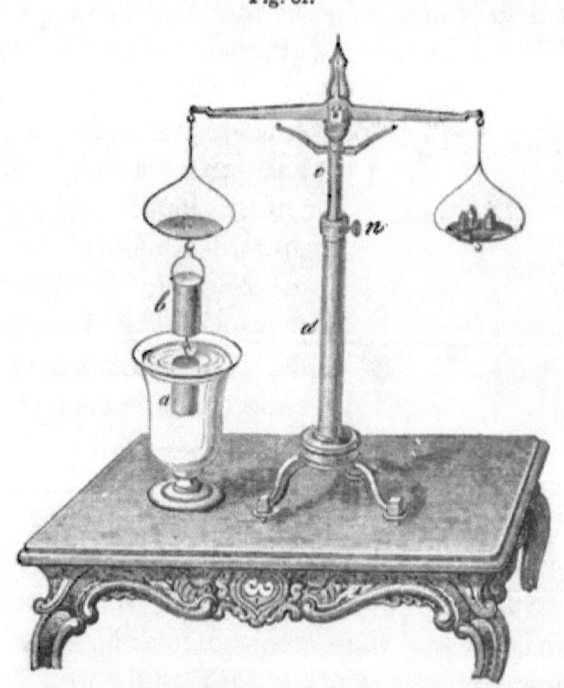

Let water be cautiously dropped into the bucket; when it is exactly full, the scales will balance again. This proves that *a body in water is buoyed up by a force equal to the weight of the water it displaces.*

To find the specific gravity of a solid body by a hy-

drostatic balance.—Weigh the body in air, and in water; the difference is the weight of its bulk of water: divide its weight in air by its loss of weight in water; the quotient is the specific gravity. Thus, sulphur loses one-half its weight when immersed in water; hence it is twice as heavy as water, and its specific gravity = 2.

To find the specific gravity of a liquid by the specific-gravity flask.—This is a bottle which holds exactly 1,000 grains of water. If it will hold 1,840 grains of sulphuric acid, the specific gravity of the acid is 1.84; if it will hold 13,500 grains of mercury, the specific gravity of that metal is 13.5.

To find the specific gravity of a liquid by a hydrometer.—This instrument consists of a glass tube, closed at one end and having at the other a bulb containing mercury or shot. A graduated scale is marked upon the tube. The *alcoholmeter* is so balanced as to sink in pure water to the zero point at the bottom of the scale. As alcohol is lighter than water, the instrument will descend for every addition of spirits which is made. The degrees of the scale indicate the percentage of alcohol. Instruments made in a similar manner are used for determining the strength of milk, acids, and solutions of various kinds.

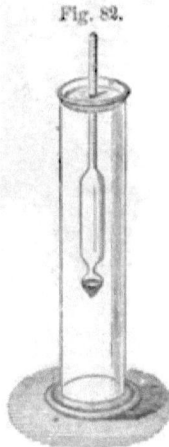

Fig. 82.

To find the weight of a given bulk of any substance.—

Multiply the weight of one cubic foot of water by the specific gravity of the substance, and that product by the number of cubic feet. Ex.: What is the weight of three cubic feet of cork? *Solution:* 1,000 oz. × .240* = 240 oz.; 240 oz. × 3 = 720 oz.

To find the bulk of a given weight of any substance.—Multiply the weight of a cubic foot of water by the specific gravity of the substance, and divide the given weight by that product. The quotient is the required bulk in cubic feet. Ex.: What is the bulk of 20,000 oz. of lead? *Solution:* 1,000 oz. × 11.36* = 11,360; 20,000 ÷ 11,360 = 1.76 + cu. ft

To find the volume of a body.—Weigh it in water. The loss of weight is the weight of the displaced water. Then, as a cubic foot of water weighs 1,000 oz., we can easily find the bulk of water displaced. Ex.: A body loses 10 oz. on being weighed in water. The displaced water weighs 10 oz. and is $\frac{1}{100}$ of a cubic foot; this is the exact volume of the body.

FLOATING BODIES.—A very pretty experiment il-

---

* TABLE OF SPECIFIC GRAVITY. (See Rev. Chem., p. 288.)

Iridium,	21.80	Flint Glass,	2.76	*Liquids.*	
Platinum,	21.53	Marble,	2.70	Sulphuric Acid,	1.84
Gold,	19.34	Quartz,	2.65	Water from the Dead	
Mercury,	13.5	Chalk,	2.65	Sea,	1.24
Lead,	11.36	Sulphur,	2.00	Milk,	1.03
Silver,	10.5	Bone,	1.99	Sea-water,	1.03
Copper,	8.9	Phosphorus,	1.83	Water,	1.
Tin,	7.3	Sugar,	1.60	Absolute Alcohol,	.79
Steel,	7.81	Coal,	1.30	Ether,	.72
Iron,	7.80	Wax,	.97		
Cast-Iron,	7.21	Ice,	.93		
Zinc,	7.	Potassium,	.86		
Heavy Spar,	4.43	Pine Wood,	.66		
Diamond,	3.50	Cork,	.24		

lustrative of this subject is represented in the cut. A glass jar is half full of water. An egg dropped in it sinks directly to the bottom. If, however, by means of a funnel with a long tube, we pour a little brine to the bottom beneath the fresh water, the egg will gradually rise. We may vary the experiment by not dropping in the egg until we have half filled the jar with the brine. The egg will then fall to the centre, and there float like a balloon. Any solid substance dissolved in water simply fills the pores of the water without adding to its bulk. This increases its density and buoyant power. A person can therefore swim much more easily in salt than in fresh water. Bayard Taylor says that he could float on the surface of the Dead Sea, with a log of wood for a pillow, as comfortably as if lying on a spring mattress. Another traveller remarks, that on plunging in he was thrown out again like a cork; and that on emerging and drying himself, the crystals of salt which covered his body made him resemble an "animated stick of rock-candy."

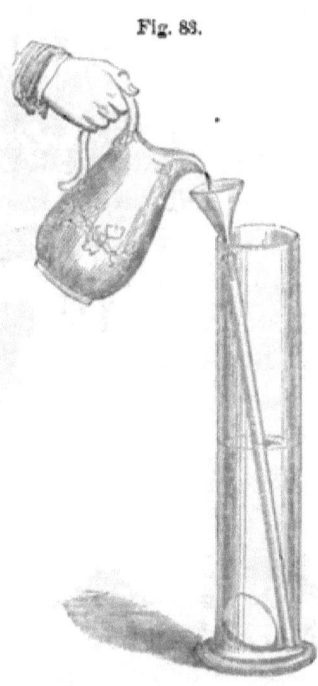

Fig. 83.

HYDROSTATICS.

A piece of iron will float, if we hammer it into a vessel so that the **weight of the water** which it displaces will exceed its own weight. An iron ship will not only float itself, but also carry a **heavy cargo**, because it displaces a great bulk of water.

A body floating in water has its centre of gravity at the lowest point. Herschel tells an amusing story of a man who attempted to walk on water by means of bulky cork boots. Scarcely, however, had he ventured out ere the law of gravitation seized him, and all that could be seen was a pair of heels, whose movements manifested a great state of uneasiness in the human appendage below.

Fish are provided with an air-bladder, placed near the spine, by means of which they can rise or sink at pleasure.

Practical Questions.—**1.** Why do housekeepers test the strength of lye, by trying whether or not an egg will float on it? **2.** How much water will it take to make a gallon of strong brine? **3.** Why can a fat man swim easier than a lean one? **4.** Why does the firing of a cannon sometimes bring to the **surface the body** of a drowned person? *Ans.* Because by the concussion it **shakes the body** loose from the mud or any object with which it is entangled. **5.** Why **does the** body of a drowned person generally come to the surface of the water, after **a time?** *Ans.* Because the gases which are generated by decomposition in the **body** render it lighter. **6.** If we let bubbles of air pass up through a jar of water, why will they become larger as they ascend? **7.** What is the pressure on a lock gate 14 feet high and 10 feet wide, when the lock is full of water? **8.** Will a pail of water weigh any more with a live fish in it than without? **9.** If the water filtering down through a rock should collect in a crevice an inch square and 250 feet high, opening at the bottom into a closed fissure having 20 square feet of surface, what would be the total pressure tending to burst the rock? **10.** Why can stones in water be moved so much more easily than on land? **11.** Why is it so difficult to wade in the water **when** there is any current? **12.** Why is a mill-dam or a canal embankment small **at the** top and large at the bottom? **13.** In digging canals and building railroads, ought not the engineer to take into consideration the curvature of the earth? **14.** Is the water at the bottom of the ocean denser than that at the surface? **15.** Why does the bubble of air in a spirit-level move as the instrument is turned? **16.** Cannot a swimmer tread on pieces of glass and other sharp substances at the bottom of the water without harm? **17.** Will a vessel draw more water in fresh or in salt water? **18.** Will iron sink in mercury? **19.** The water in the reservoir in New York is about 80 feet above **the fountain** in the City Hall Park. What is the pressure upon a single inch of the pipe at the latter point? **20.** Why does cream rise on milk? **21.** If a ship founders at sea, to what depth will she descend? (* It is a poetical thought that ships may thus sink into submarine currents and be **carried** hither and thither with their precious cargoes of freight and passengers, on voyages that know no end and toward harbors that they never reach.) **22.** There is a story told of a Chinese boy who accidentally dropped his ball into a deep hole where he could not reach it. He filled the hole with water, but the ball would not quite float. He finally bethought himself of a lucky expedient, which was successful. Can you guess it? **23.** Which has the greater buoyant force, water or oil? **24.** What is the weight of four cubic feet of cork? **25.** How many ounces of iron will a cubic foot of cork float in water? **26.** What is the specific gravity of a body whose weight in air is 30 grs. and in water 20 grs.? How much is it heavier than water? **27.** Which is heavier, a pail of fresh or one of salt water? **28.** The weights of a piece of syenite-rock in air and water were 3941.8 grs. and 2607.5 grs. Find its specific gravity. **29.** A specimen of green sapphire from **Siam** weighed in air 21.45 grs. and in water 16.33 grs.; required its specific **gravity. 30.** A specimen of granite weighs in air 534.8 grs. **and in water 334.6 grs.**; what is its specific gravity? **31.** What is the bulk of **a ton of** iron? A ton of gold? A ton of copper? **32.** What is the weight of a cube of gold **4 feet on** each side? **33.** A cistern is 12 feet long, 6 feet wide, and 10 feet deep; when full of water, what is the pressure on each side? **34.** Why does a dead fish always float on its back? **35.** A given bulk of water weighs 62.5 grs., and the same bulk of muriatic acid 75 grs. What is the specific gravity of the acid? *Ans.* 1.2. **36.** A vessel holds 10 lbs. of water; how much mercury would it contain? **37.** A stone weighs 70 lbs. in air and 50 in water; what is its bulk? **38.** A hollow **ball of iron** weighs 10 lbs.; what must be its bulk to float in water?

HYDRAULICS.

HYDRAULICS treats of liquids in motion. In this, as in Hydrostatics, water is taken as the type. In theory, its principles are those of falling bodies, but they are so modified by various causes, that in practice they cannot be relied upon except as verified by experiment. The discrepancy arises from changes of temperature which vary the fluidity of the liquid, from friction, the shape of the orifice, &c.

The velocity of a jet is the same as that of a body falling from the surface of the water.—We can see that this must be so, if we recall two principles we have already learned. First, that "**a jet will rise to the level of its source;**" and second, that "**to elevate a body to any height, it must have the same velocity that it would** acquire in falling that distance." It follows, therefore, that the velocity of a jet depends entirely on the height of the liquid above the orifice, and that all liquids will issue with the same velocity at the same depth. Molasses ought to flow with the same speed as mercury, for the same reason that a guinea falls in the same time as a feather. The application of this principle is of course modified by the temperature, and various other causes.

To find the velocity of a jet of water.—We use here the 4th equation of falling bodies, $v = 2\sqrt{gd}$, in which d is the distance of the orifice below the surface of the water. Ex.: The depth of water above the orifice is 64 feet; required the velocity. Substituting 64 for d, we have $v = 2\sqrt{16 \times 64} = 64$ feet.

To find the quantity of water discharged in a given time.—Multiply the area of the orifice by the velocity of the water, and that product by the number of seconds. Ex.: What quantity of water will be discharged in five seconds from an orifice having an area of $\frac{1}{3}$ a square foot, at a depth of 16 feet? At that depth, $v = 2\sqrt{16 \times 16} = 32$ feet per second; multiplying by ½, we have 16 cubic feet as the amount discharged in one second and 80 cubic feet in five seconds. In practice, however, it is found that but 62 per cent. of this amount can be realized.

EFFECT OF TUBES.—If we examine a jet of water, we shall see, just outside the orifice, its size is decreased to about $\frac{2}{3}$ that at the opening. This is caused by the water producing cross currents as it flows from different directions toward the orifice. If a tube of a length twice or thrice the diameter of the opening be inserted, the water adheres to the sides of the tube, so that there is no contraction, and the flow is increased to 82 per cent. of the theoretical amount.

If the tube be conical, and inserted with the large end in the opening, the discharge may be increased to 92 per cent.; and strangely enough, by inserting it with the smaller end next the orifice, the amount exceeds that indicated by theory as much as 25 per cent. It seems in this case to be made so easy for the water to run, that more is coaxed out than ought to go. Long tubes or curves, however, by their friction, largely diminish the flow of water. It is said that a single right-angle will decrease it one-half, while an inch pipe 200 feet long will discharge only ½ as much water as one an inch long.

FLOW OF WATER IN RIVERS.—A fall of only three inches per mile is sufficient to give motion to water, and produce a velocity of as many miles per hour. The Ganges descends but 800 feet in 1,800 miles. Its waters require a month to move down this long inclined plane. A fall of 3 feet per mile will make a mountain torrent. The current moves more swiftly at the centre than near the shores or bottom of a channel, since there is less friction.

WATER-WHEELS are machines for using the force of falling water. By means of bands or cog-wheels the motion of the wheel is conducted from the axle into the mill. The principle is that of a lever with the P acting on the short arm. In this way, the movement of the slow creaking axle reappears in the swiftly buzzing saw or flying spindle. Water-wheels are of four classes— *The Overshot, Undershot, Breast,* and *Turbine* wheels. *The Overshot-wheel* has on its circumference a series of buckets which receive the water as it flows out of a *sluice*, C. The

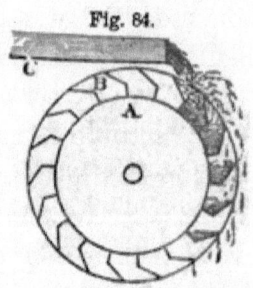

Fig. 84.

buckets are so made as to hold the water as they descend on one side, and to empty it as they come up on the other. Overshot-wheels are valuable where a great fall can be secured, since they require but little water. They are made of great size. One at Cohoes, N. Y., is 96 feet high. If P denotes the weight of the water and d the distance it falls, then the total force $= Pd$. Of this amount 80 per cent. can be secured in the best wheels of this and the third class.

The Undershot-wheel, instead of buckets has merely projecting boards or *floats*, which receive the force of the current. It is of use where there is little fall and a large quantity of water. It is said to utilize only 20 per cent. of the force of the water.

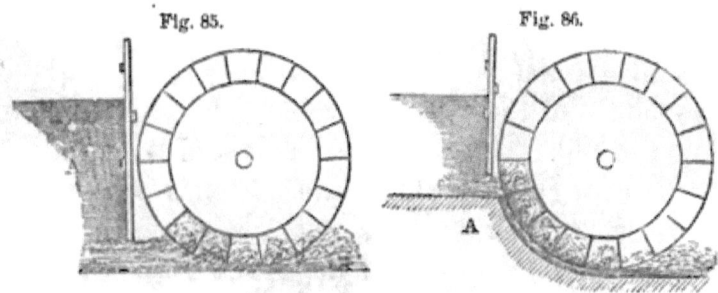

Fig. 85.　　　　　Fig. 86.

The Breast-wheel is a medium between the two before named, as may be seen in Fig. 86.

The Turbine-wheel differs essentially from the others named. It is placed horizontally, and is entirely immersed in the water. In the figure, C is the dam and D A the spout by which the water is furnished to the wheel. E is a scroll-like casing encircling the wheel, and open at the centre above and below. The axis of the wheel is the cylinder f, from which radiate plane-floats against which the water strikes. To confine the water at the top and the bottom is a circular disk attached to the cylinder and the floats. In these disks are the swells for discharging the water. They project above and below, as seen in the figure. They commence near the cylinder, and swelling outward scroll-shaped, form openings curved toward the cylinder, thus emptying the water in a direction opposite to that in which it enters the wheel. This form utilizes as high as 90 per cent. of the force. F is a band-wheel which conducts the power to the machinery. The principle of the turbine is

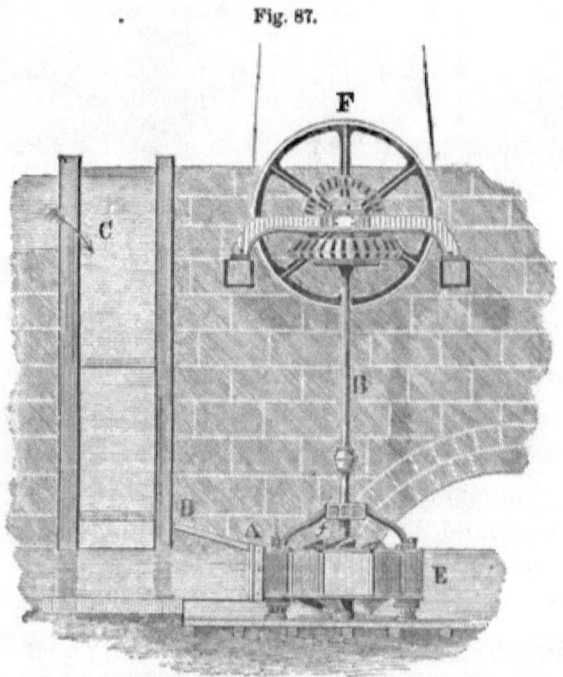

Fig. 87.

that of the unbalanced pressure of a column of water. It is finely illustrated in the old-fashioned Barker's Mill or Reaction Wheel. This consists of an upright cylinder with horizontal arms, on the opposite sides of which are small apertures. It rests in a socket, so as to revolve freely. Water is supplied from a tank above. If the openings in the arms are closed, when the cylinder is filled with water the pressure will be equal in all directions and the machine will be at rest. If now we open an aperture, the pressure is relieved on that side, and the arm flies backward with the unbalanced pressure of the column of water above.

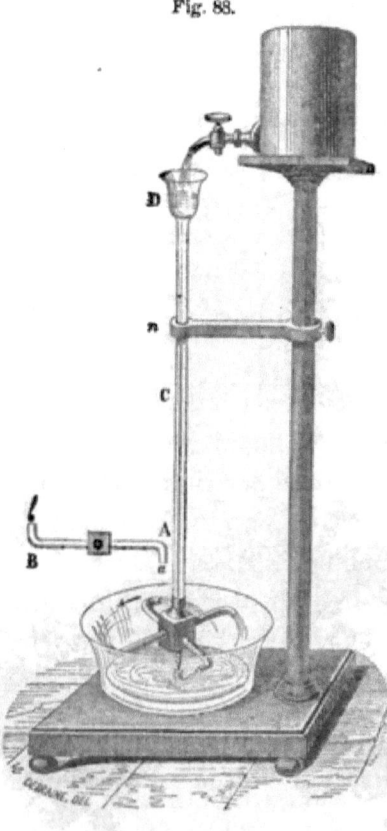

Fig. 88.

WAVES are produced by the friction of the wind against the surface of the water. A light wind forms merely ripples; these increase out in the open sea, as wave is raised upon wave, until they become great billows which constantly surge to and fro, so that the sea is never at rest. The wind raises the particles of water and gravity draws them back again. They thus vibrate up and down, but do not advance. The forward movement of the wave is only an illusion. The *form* of the wave progresses, but not the water of which it is composed, any more than the thread of a screw which we turn in our hand; or the undulations of a rope or carpet which is being shaken; or the stalks of grain which bend in billows as the wind sweeps over them. If we watch a buoy in the harbor or a body floating on the surface of water, we shall see that it moves forward on the crest of each wave through a few feet or inches, according to the length of the wave; then stops, moves backward in the hollow; stops, and again moves forward as before on the crest of the next wave. The molecules of water vibrate to and fro in an elliptical path. Thus, let the figure represent two successive wave-crests and the hollow between.

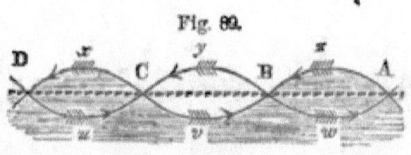

Fig. 89.

While the whole wave moves from the position A B to that of C D, the molecules of water only move backward or forward through a distance A B or C D; forward on the crest of the wave and backward in the hollow as shown by the arrows. The velocity of the particles may be much slower than that of the progressive motion of the wave. It is said that in an earthquake, the velocity of the particles of the shaken ground is often only three feet per second, while the earth-wave moves across the country at the rate of 3,000 feet per second.

Near the shore the character of waves is somewhat changed. The oscillations are shorter, and as the waves do not balance those in the deeper water, they are forced forward till the lower part of each wave is checked by the friction on the sandy beach, and the upper part curls over and falls beyond. The size of "mountain billows" has been much exaggerated. The ocean is probably undisturbed below the depth of 30 feet. The highest wave, from the deepest "trough" to the very topmost "crest," is only 43 feet. The corresponding parts of different waves are termed *like phases*. The distance between two like phases, or between the crests of two succeeding waves, is called a *wave-length*. *Opposite phases* are those parts which are vibrating in different directions, as the point midway in the front of one wave and another midway in the rear of the next wave.

A tide-wave may be setting steadily toward the west; waves from distant storms may be moving upon this; and above all, ripples from the breeze then blowing may diversify the surface. These different systems will each be entirely distinct, yet the joint effect may be very peculiar. If any two systems exactly coincide with *like phases*,—the crest of one meeting the crest of the other, and the furrow of one meeting the furrow of the other,—the resulting wave will have a height equal to the *sum of the two*. If any two coincide with opposite phases,—the hollow of one striking the crest of another,—the height will be the *difference of the two*. Thus, if in two systems having the same wave-length and height, one is exactly half a length behind the other, they will mutually destroy each other. This is termed the *interference* of waves. The manner in which different waves move among and upon each other, is seen by dropping a handful of stones in water and watching the waves as they

Fig. 90.

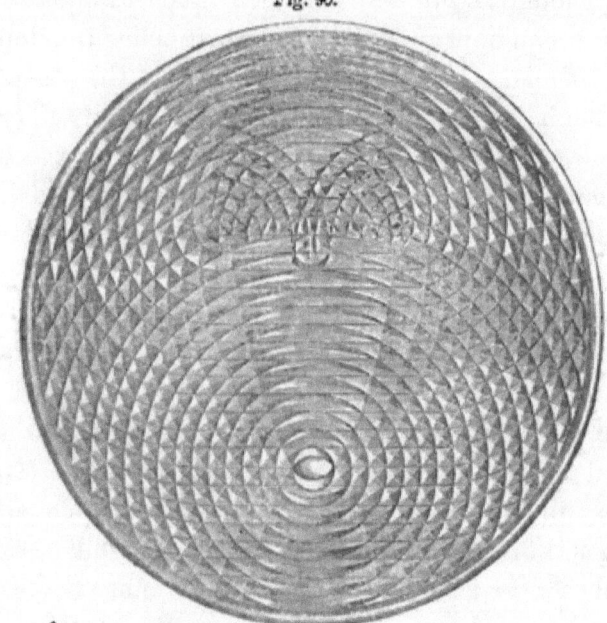

circle out from the various centres in ever-widening curves. In the figure is shown the beautiful appearance these waves present when reflected from the sides of a vessel.

The application of these principles in Sound and Light will be found very important.

Practical Questions.—1. How much more water can be drawn from a faucet 8 feet than from one 4 feet below the surface of the water in a cistern? 2. How much water will be discharged per second from a short pipe having a diameter of 4 inches and a depth of 48 feet below the surface of the water? 3. When we pour molasses from a jug, why is the stream so much larger near the nozzle than at some distance from it? 4. Ought a faucet to extend into a barrel beyond the staves? 5. What would be the effect if both openings in one of the arms of Barker's Mill were on the same side?

PNEUMATICS.

PNEUMATICS treats of the general properties and the pressure of gases. Since the molecules move among each other more freely even than those of liquids, the conclusions at which we have arrived with regard to *transmission of pressure, buoyancy and specific gravity* apply also to gases. Its principles obtain in all gaseous bodies, but as air is the most abundant gas, it is taken as the type of the class, as water is of liquids.

THE AIR-PUMP is shown in its essential features in Fig. 91. A is a glass receiver standing on an oiled pump-plate. The tube D, connecting the receiver with the cylinder, is closed by the valve E opening upward. There is a second valve, P, in the piston, also opening upward. Suppose the piston is at the bottom and both valves shut. Let it now be raised, and there will be a vacuum produced in the cylinder; the expansive force of the atmosphere in the receiver will open the valve E and drive the air through to fill this empty space. When the piston descends, the valve E will close, while the valve P will open, and the air will pass up above the piston. On elevating the piston a second time, this air is removed

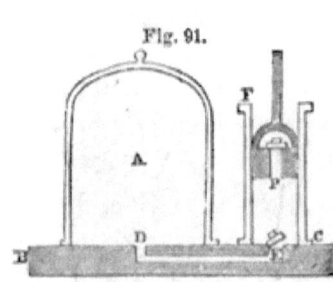
Fig. 91.

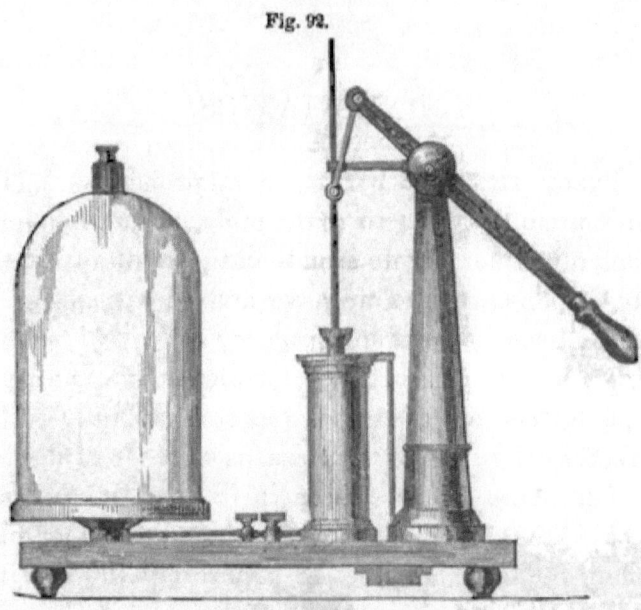

Fig. 92.

The Air-pump.

from the cylinder, while the air from the receiver passes through as before. At each stroke a portion of the atmosphere is drawn off; but the expansive force becomes less and less, until finally it is not sufficient even to lift the delicate valves. For this reason a perfect vacuum cannot be obtained.

PROPERTIES OF THE AIR.—*Weight.*— Exhaust the air from a flask which holds 100 cubic inches, and then balance it accurately. If now we turn the stop-cock, the air will rush in with

Fig. 93.

a whizzing noise and the flask will descend. We shall have to add about 31 grains to restore the equipoise.

Elasticity and compressibility.—These properties are shown in the common pop-gun. We compress the atmosphere in the barrel until the elastic force becomes so great as to drive out the stopper with a loud report. As we crowd down the piston we feel the elasticity of the air yielding to our strength, like a cushion or a bent spring.

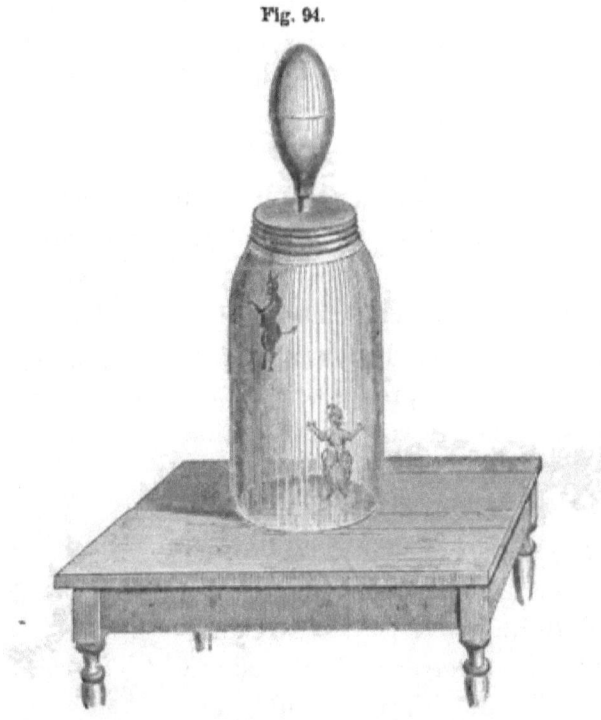

Fig. 94.

The bottle-imps, or Cartesian divers, illustrate the same properties. Fig. 94 represents a very simple form of this experiment. The cover of a common fruit-jar is fitted with a small tin tube, which is inserted into a syringe-bulb. The jar is filled with water and the divers placed within. These are hollow images of glass, having each a small opening at the end of the curved tail. If we squeeze the bulb, the air will be forced into the jar and the water will transmit the pressure to the air in the image. This being compressed, the water will enter, and the specific gravity being increased, the diver will descend. On relaxing the grasp of the hand on the bulb, the air will return into it, the air in the image will expand by its elastic force driving out the water, and the diver, thus lightened of his ballast, will ascend. The nearer the image is to the bottom, the less force will be required to move it. With a little care it can be made to respond to the slightest pressure, and will rise and fall as if instinct with life. This experiment shows also the buoyant force of liquids, their transmission of pressure in every direction, the increase of the pressure in proportion to the depth, and the principle of Barker's Mill.

Expansibility.—Let a well-dried bladder be partly filled with air and tightly closed. Now place it under the receiver and exhaust the air. The air within the bladder expanding will swell and oftentimes burst it into shreds.

Take two bottles partly filled with colored water.

Let a bent tube be inserted tightly in A and loosely in B. Place this apparatus under the receiver and exhaust the air. The expansive force of

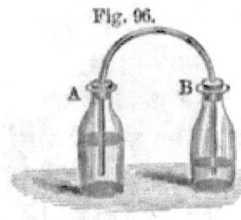

Fig. 96.

the air in A will drive the water over into B. On readmitting the air into the receiver, the pressure will return the water into A. It may thus be driven from one bottle to the other at pleasure.

PRESSURE OF THE AIR.—If we place the handglass on the plate of the Air-pump, covering it with

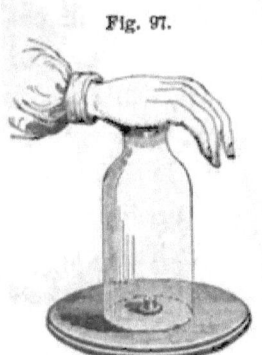

Fig. 97.

one hand, on exhausting the air we shall soon find the pressure to become painful. Tie over one end of the glass a piece of well-soaked bladder. When thoroughly dry, exhaust the air from it as before, and the membrane will burst with a sharp report.

The Magdeburg Hemispheres are named from the city in which Otto Guericke, their inventor, resided. They consist of two small brass hemispheres, which fit closely

Fig. 98.

together, but may be separated at pleasure. If, however, the air be exhausted from within, the strength of several persons will be required to pull them apart. No matter in what position the hemispheres are held, we shall find the pressure the same.

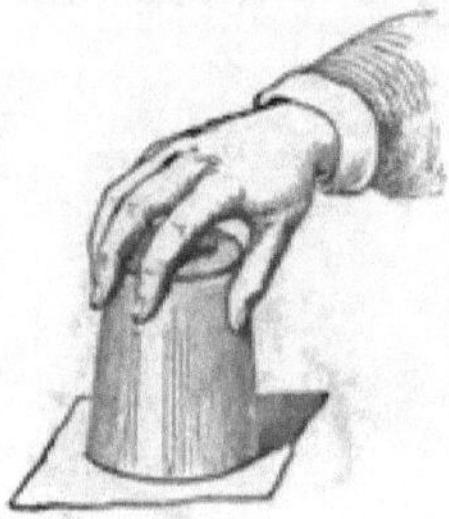

Upward Pressure of the Air.—Fill a tumbler with water, and then lay a sheet of paper over the top. Quickly invert the glass, and the water will be supported by the upward pressure of the air.

Fig. 99.

Within the glass cylinder shown in Fig. 100 there is a piston working airtight. Connect C with the pump by means of a rubber tube and exhaust the air. The weight will leap up as if caught by a spring.

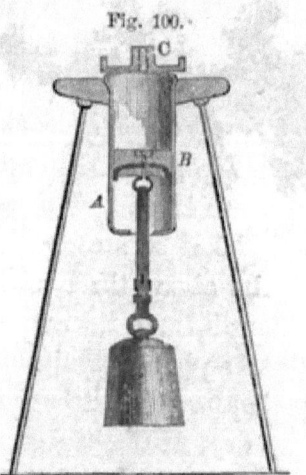

Fig. 100.

Buoyant Force of the Air.—The principle of Archimedes holds true in gases as in liquids. Illustrations of this abound in common life. Smoke and other light substances float in the air, as wood does in water, because they are lighter and are buoyed with a force equal to the weight of the air they displace. In Fig. 101 we have a hollow sphere of copper, which is exactly balanced in the air by a solid lead weight, but instantly falls on being placed under the receiver and the air exhausted. This shows that its weight was partly sustained by the buoyant force of the air.

The pressure of the air sustains a column of mercury

Fig. 101.

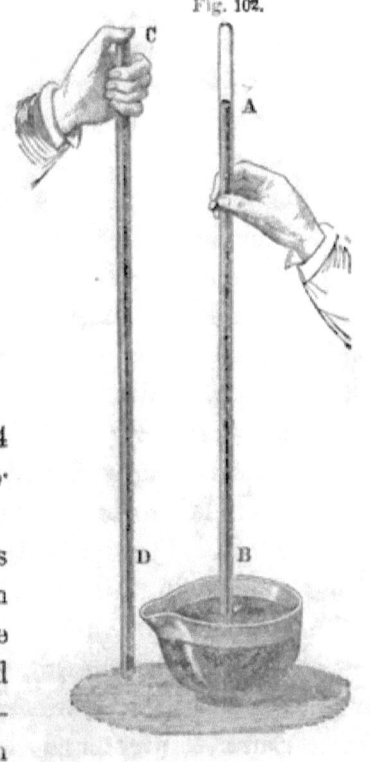

Fig. 102.

30 inches high, of water 34 feet high, and is 15 lbs. per square inch.

Take a strong glass tube about three feet in length, and tie over one end a piece of well-soaked bladder. When thoroughly dry, fill the tube with mercury, and invert it in a cup of the same liquid. The mercury will sink to a height of about 30 inches. If the area of the tube be one inch, this amount of the metal will weigh about 15 lbs. The weight of the column of mercury is equal to the downward pressure on each square inch of the surface of the mercury in the cup. Hence we conclude that the pressure of the atmosphere is 15 lbs. per square inch, and will balance a column of mercury 30 inches high. As water is 13½ times lighter than mercury, it is evident that the same pressure would balance a column of that liquid 13½ times higher, or 33¾ feet. On account of the unwieldy length of the tube required to exhibit the column of water, it is not easy to verify this last statement. It may, however, be prettily illustrated in the following manner. Pour on the mercury in the cup (Fig. 102) a little water colored with red ink. Now raise the end of the tube carefully above the surface of the metal, but not above that of the water which will immediately rise in the tube, the mercury passing down in beautifully beaded globules. The mercurial column was only 30 inches high, while the water will entirely fill the tube. Finish the experiment by puncturing the bladder with a pin, when the water will instantly fall to the cup below.

The pressure of the air varies.—We live on the bed of an aërial ocean whose invisible tides surge around us on every side. More restless than the sea, its waves beat to and fro, stirred by a multitude of causes. Changes of temperature, moisture, &c., constantly vary the weight of the air, and consequently change the height of the column of liquid which it can support. There is also a diurnal variation, due to the heat of the sun,—slight indeed, yet so marked that Humboldt says that the play of the mercurial column could be used to indicate the hour of the day. The pressure of the air increases with the depth. Hence, in a valley its weight is greater than on a mountain.

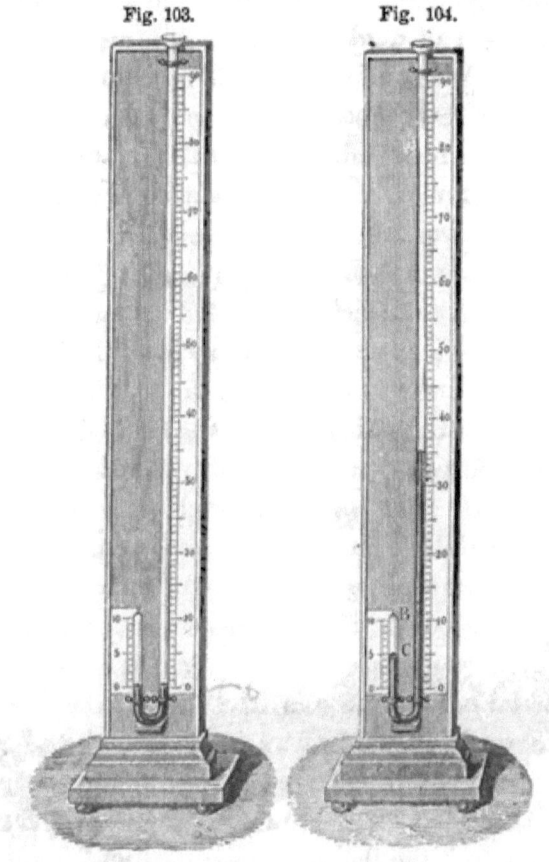

Fig. 103. Fig. 104.

The figures given in the last paragraph apply only to the level of the sea and the temperature of 60° F. They are an average of all the variations, and are considered the standard for reference.

Mariotte's Law.—Fig. 103 represents a long bent glass tube with the end of the short arm closed. Pour mercury into the long arm until it rises to the point marked zero. It stands at the same height in both arms, and there is an equilibrium. The air presses on the mercury in the long arm with a force equal to a column of mercury 30 inches high, and the elastic force of the air confined in the short arm is equal to the same amount. Let us now pour additional mercury into the long arm until it stands at 30 inches. (Fig. 104.) We have evidently doubled the pressure. If we look at the short arm, we shall find that the air is condensed to one-half its former dimensions, and of course the expansive force must be doubled. We therefore conclude that *the elasticity of a gas increases and the volume diminishes in proportion to the pressure upon it.*

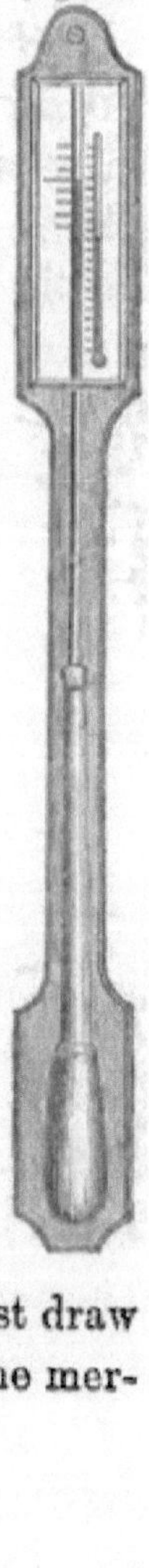

The Barometer is an instrument for measuring the pressure of the air. It consists essentially of the tube and cup of mercury shown in Fig. 102. A scale is attached for convenience of reference. The barometer is used (1) to indicate the weather, and (2) to measure the height of mountains.

It does not absolutely foretell the character of the weather. It simply shows the varying weight of the air, from which we must draw our own conclusions. A continued rise of the mer-

Fig. 105.

cury indicates fair weather, and a continued fall, foul weather.

Since the pressure diminishes as one ascends above the level of the sea, the observer ascertains the fall of the mercury in the barometer, and the temperature by the attached thermometer; and then, by reference to carefully prepared tables, easily determines the height.

Water-barometer.—Mercury is used for filling the barometer because of its weight and its low freezing-point. Water would require a tube about 34 feet in length. It is said that the first barometer was filled with that liquid. The inventor, Otto Guericke, a wealthy burgomaster of Magdeburg, Saxony, erected a tall tube reaching from a cistern in the cellar up through the roof of his house. A tall wooden image—life-size—was placed within the tube, floating upon the water. On fine days, this novel weather-prophet would rise above the roof-top and peep out upon the queer old gables of that ancient city, while in foul weather he would retire to the protection of the garret. The accuracy of these movements attracted the attention of the neighbors. Finally, in their innocency, becoming suspicious of Otto Guericke's piety, they openly accused him of being in league with the devil. So the offending philosopher relieved this wicked wooden man from longer dancing attendance upon the weather, and the staid old city was once more at peace.

Pumps.—Two varieties are in common use. These are the *Lifting* and the *Forcing* pump.

The Lifting-pump contains two valves opening upward—one, *a*, at the top of the *suction-pipe*, B; the other, *c*, in the piston. Suppose the handle to be raised, the piston at the bottom of the cylinder

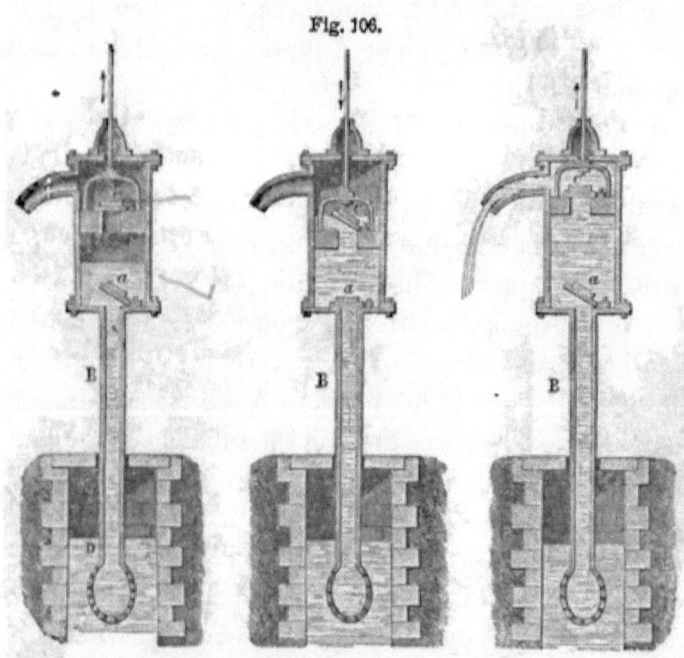

Fig. 106.

and both valves closed. Now depress the pump-handle and elevate the piston. This will produce a partial vacuum in the suction-pipe. The pressure of the air on the surface of the water below will force the water up the pipe, open the valve, and fill the chamber, as seen in the first figure. Let the pump-handle be elevated again, and the piston depressed. The valve *a* will now close, the valve *c* will open and the water will flow through it above the piston, as in the second figure. When the pump-handle is lowered the second time and the piston elevated, the water is lifted up to the spout, whence it flows out; while at the same time the lower valve opens and the water is forced up from below by the pressure of the air, as in the third figure.

If the valves and piston were fitted air-tight, the water could be raised 34 feet (more exactly 13½ times the height of the barometric column) to the lower valve, but owing to various imperfections it commonly reaches only 28 feet. For a similar reason we sometimes find a dozen strokes necessary to "bring water."

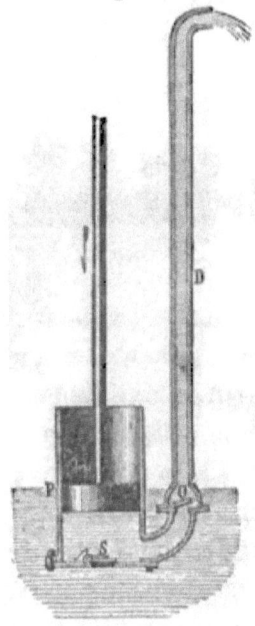

The Force-pump has no valve in the piston. The water rises above the lower valve as in the lifting-pump. When the piston descends, the pressure opens the valve O and forces the water up the pipe D. This pipe may be made of any length, and thus the water driven to any height.

The Fire-engine consists of two force-pumps with an air-chamber. The water is driven by the pistons m, n, alternately into the chamber R, whence the air, by its ex-

PNEUMATICS.   145

pansive force, throws it out in a continuous stream through the hose-pipe attached at Z.

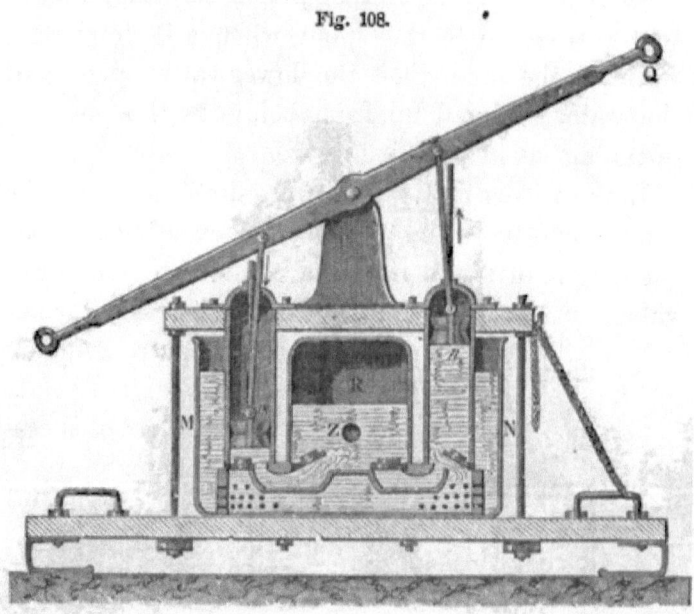

Fig. 108.

THE SIPHON consists simply of a tube bent in the form of the letter U, having one arm longer than the other. We insert the short arm in the water, and then applying the mouth to the other, exhaust the air. The water will immediately begin to flow from the long arm, and continue until the lower end of the short arm is uncovered, or until the water in the two vessels comes to the same level. A very instructive variation of this experiment may be given if we color the water with red ink, and then allow it to run from one tumbler into another until just before the flow would cease; then quickly elevate the vessel containing the long arm, carefully keeping both ends of the siphon under the water, when the flow will set back to the first vessel. Thus we may alternate,

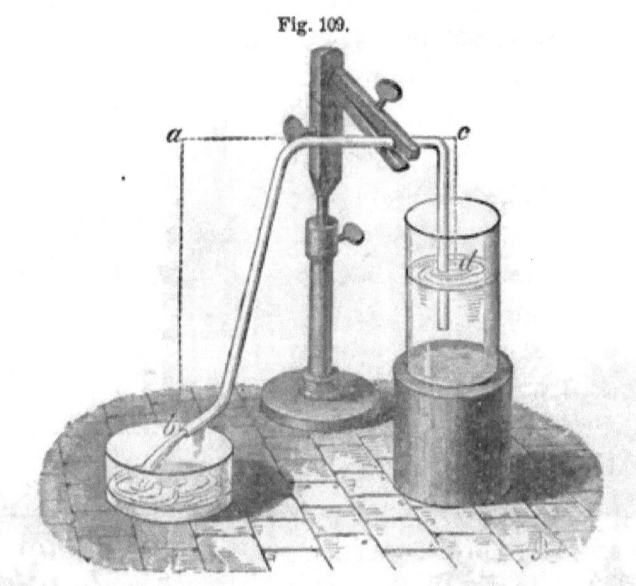

Fig. 109.

backward and forward, until we see clearly that the water flows always to the lower level from the long arm, and ceases whenever the water in the two vessels reaches the same level.

The Theory of the Siphon.—The pressure of the air at b holds up the column of water $a\,b$, and the upward pressure is the weight of the air less the weight the column of water $a\,b$. The upward pressure at d is the weight of the air minus the weight of the column of water $c\,d$. Now $c\,d$ is less than $a\,b$, therefore the pressure at d is greater than that at b, and the water in the tube is driven toward the longer arm by a force equal to the difference between the two arms.

THE PNEUMATIC INKSTAND is filled by pouring in the ink when the bottle is tipped so that the nozzle is at the top. The pressure of the air will then hold the ink in the stand. When it is used below the level of o, a bubble of air passes in, forcing the ink into the nozzle as desired.

Fig. 110.

THE ANCIENTS noticing how the air rushes in to fill any empty space, explained the fact by saying, "Nature abhors a vacuum." This principle answered the purpose of philosophers even in modern times. In the 17th century, when workmen were employed by the Duke of Tuscany to dig a very deep well near Florence, they found to their surprise that the water would not rise in the pump as high as the lower valve. They applied in their dilemma to Galileo. The old philosopher replied— half in jest we hope, certainly he was half in earnest— "Nature does not abhor a vacuum beyond 34 feet."

THREE OPPOSING FORCES ACT UPON THE AIR—viz: Gravity, which binds it to the earth, and the Centrifugal and the Repellant [heat] forces, which tend to hurl it off into space. Under the action of the latter forces the atmosphere, **like a great** bent spring, is ready to bound away at the first opportunity; but the attraction of the earth holds it firmly in its place.

HEIGHT AND DENSITY OF THE AIR.—Fifty miles has been taken as the extreme limit of the atmosphere. The latest investigations, however, indicate that there is an extremely rarefied air at the height of perhaps 500 miles. Its density rapidly diminishes as we ascend. At the height of $3\frac{1}{2}$ miles it is but one-half that at the sea-level. At 40 miles the atmosphere is rare **as in the vacuum of** an air-pump.

Practical Questions.—**1.** Why must we make two openings in a barrel of cider when we tap it? **2.** What is the weight of 10 cubic feet of air? **3.** What is the pressure of the air on 1 square rod of land? **4.** What is the pressure on a pair of Magdeburg hemispheres 4 inches in diameter? **5.** How high a column of water can the air sustain when the barometric column stands at 28 inches? **6.** If we should add a pressure of two atmospheres (30 lbs. to the square inch), what would be the volume of 100 cubic inches of common air? **7.** If, while the water is running through the siphon, we quickly lift the long arm, what is the effect on the water in the siphon? If we lift the entire siphon? **8.** When the mercury stands at $29\frac{1}{2}$ inches in the barometer, how high above the surface of the water can we place the lower pump-valve? **9.** Why cannot we raise water by means of a siphon to a higher level? **10.** If the air in the chamber of a fire-engine be condensed to $1/_{16}$ its former bulk, what will be the pressure due to the expansive force of the air on every square inch of the air-chamber? **11.** What causes the bubbles to rise to the surface when we put a lump of loaf-sugar in hot tea? **12.** To what height can a balloon ascend? What weight can it lift? **13.** The rise and fall of the barometric column shows that the air is lighter in foul and heavier in fair weather. Why is this? *Ans.* In fair weather the moisture of the air is an invisible vapor mingled with it and adding to its pressure, while in foul weather the vapor is separated in the form of clouds. **14.** When smoke ascends in a straight line from chimneys, is it a proof of the rarity or density of the air? **15.** Why do we not feel the heavy pressure of the air upon our bodies? **16.** Is a bottle empty when filled with air? **17.** Why is it so tiresome to walk in miry clay? *Ans.* Because the upward pressure of the air is removed from our feet. **18.** How does the variation in the pressure of the air affect those who ascend lofty mountains? Who descend in diving-bells? **19.** Explain the theory of "sucking cider" through a straw.

On Sound.

"Science ought to teach us to see the invisible as well as the visible in nature: to picture to our mind's eye those operations that entirely elude the eye of the body; to look at the very atoms of matter, in motion and in rest, and to follow them forth into the world of the senses."—TYNDALL.

ACOUSTICS.

Acoustics treats of sound.*

SOUND IS PRODUCED BY VIBRATIONS.—By lightly tapping a receiver or even a glass fruit-dish, you can see that the sides are thrown into motion. Fill a goblet half full of water, and wetting your finger, rub it lightly around the upper edge of the glass. The sides will vibrate, and tiny waves corresponding to these movements will ripple the surface of the water. The vibrations of a tuning-fork are very distinct. Hold a card close to its prongs, and you can hear

---

* The term *sound* is used in two senses—the *subjective* (that which has reference to our mind) and the *objective* (that which refers only to the objects around us). (1) Sound is the sensation produced upon the organ of hearing by vibrations in matter. In this use of the word there can be no sound where there is no ear to catch the vibrations. An oak falls in the forest, and if there is no ear to hear it there is no noise, and the old tree drops quietly to its resting-place. Niagara's flood poured over its rocky precipice for ages, but fell silently to the ground. There were the vibrations of earth and air, but there was no ear to receive them and translate them into sound. When, however, the first foot trod those primeval solitudes, and the ear first felt the pulsations from the torrent, then **first the roaring cataract** found a voice and broke its lasting silence. A trumpet does not sound. It only carves the air into waves. The tympanum is the beach on which these break into sound. (2) Sound is those vibrations of matter capable of producing a sensation upon the organ of hearing. In this use of the word there can be a sound in the absence of the ear. An object falls and the vibrations are produced, though there may be no organ of hearing to receive an impression from them.

the repeated taps. Place your cheek near them, and you will feel the little puffs of wind. Insert the handle between your teeth, and you will experience the indescribable thrill of the swinging metal. The tuning-fork may be made even to draw the outline of its vibrations upon a smoked-glass. Fasten upon one prong a sharp-pointed piece of metal, and on

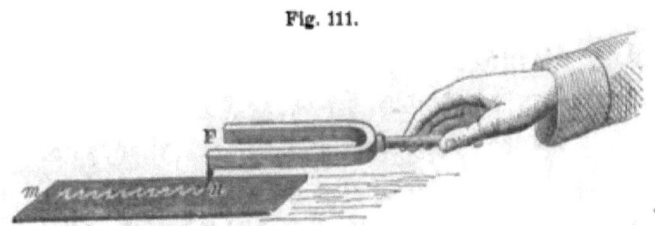

Fig. 111.

drawing the fork along as in the figure, a sinuous line will show the width (amplitude) of the vibrations.

From many similar experiments it is believed that when a body is struck its molecules are thrown into motion, and that all sound is produced by vibrations.

How sound is conveyed through the air.—Let us imagine the prong of the tuning-fork used in the last experiment to advance, condensing the air in front of it, and then to recede, leaving behind it a partial vacuum. This process is repeated until the fork comes to rest, and the sound ceases. Each vibration of the prong produces a *sound-wave* of air, which contains one condensation and one rarefaction. In water, we measure a wave-length from crest to crest; in air, from condensation to condensation. The condensation of the sound-wave corresponds to the crest of the water-wave, and the rarefaction of the sound-wave to the hollow of the water-wave. In the figure,

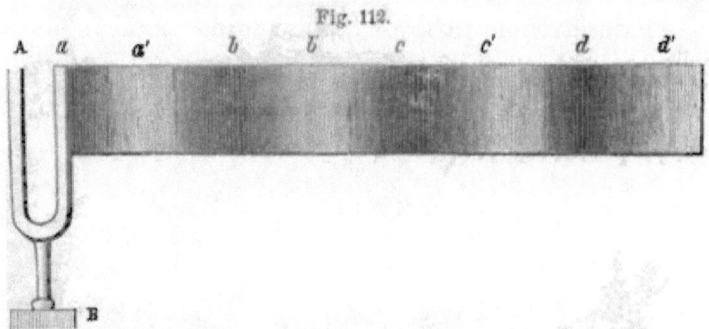
Fig. 112.

the dark spaces a, b, c, d represent the condensations, and a', b', c' the rarefactions; the wave-lengths are the distances ab, bc, cd.

If we fire a gun, the gases which are produced expand suddenly and force the air outward in every direction. This hollow shell of air thus condensed imparts its motion to that next to it, while it springs back by its elasticity and becomes rarefied. The second shell rushes forward with the motion received, then bounds back and becomes rarefied. Thus each shell of air takes up the motion and imparts it to the next. The wave, consisting of a condensation and a rarefaction, proceeds onward. It is, however, as in water-waves, a movement of the *form* only, while the particles vibrate but a short distance to and fro. The molecules in water-waves oscillate *vertically ;* those in sound-waves *horizontally*, or parallel to the line of motion.

If a bell be rung, the air adjacent to it is set in motion: thence, by a series of condensations and rarefactions, the vibrations of the bell are conveyed to the ear, and thus produce the sensation of sound.

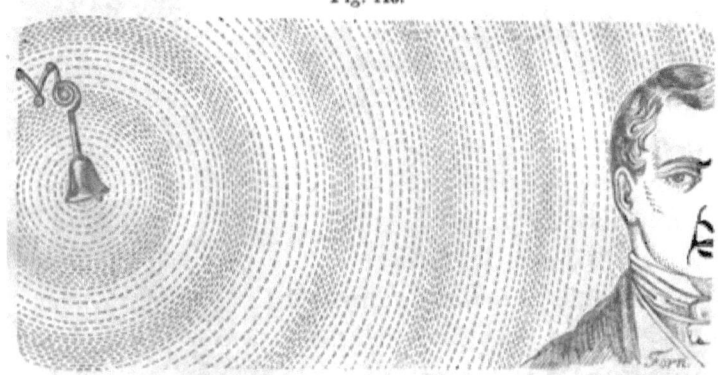

Fig. 113.

When we speak, we do not shoot the air which we expel from our lungs into the ear of the listener. We simply condense the air just before our mouth and throw it into vibrations. Thus a sound-wave is formed. This travels onward and spreads in every direction in the form of a sphere of which we are the centre.

SOUND WILL NOT PASS OUT OF A VACUUM.—In the figure, B, the bell, is struck by clock-work which may be set in motion by the sliding-rod r. The apparatus is suspended by means of silk cords, that no vibration may be conducted through the pump itself. If the air be exhausted, the sound will become so faint that it cannot be heard, even when the ear is placed close to the receiver.

There is perfect silence in a vacuum. No sound can therefore be transmitted to the earth from the regions of space. The movements of the heavenly bodies are noiseless. In the expressive language of David, "Their voice is not heard." In elevated regions sounds are diminished in loudness. The explosion of a pistol on Mont Blanc is said to resemble that of an ordinary fire-cracker; and it is difficult to continue a conversation, as the voice must be raised so far above its natural pitch. The reverse of this takes place when persons descend into deep mines or in diving-bells. The sounds then become startlingly distinct, and the workmen are compelled to talk in whispers.

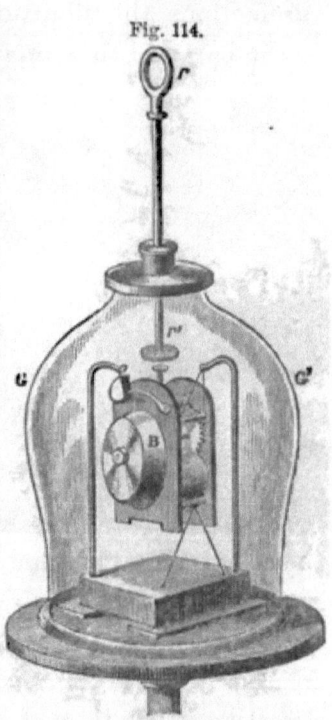

Fig. 114.

THE VELOCITY OF SOUND depends on the *elasticity* and *density* of the medium through which it passes. The higher the elasticity, the more promptly and rapidly the motion will be transmitted, since the elastic force acts like a bent spring between the molecules: the greater the density, the more molecules to be set in motion, and hence the slower the transmission.

Sound travels through the air (at the freezing-point) at the rate of 1,090 *feet per second.*—A rise in temperature diminishes the density of the air, and thus sound travels faster in warm and slower in cold air. A difference of 1° F. makes a variation of about 1 foot in velocity.

Sound travels through water at the rate of 4,700 *feet per second*—Water is denser than the air, and for that reason sound should travel in it much more slowly; but its elasticity, which is measured by the force required to compress it, is so much greater, that the rate is quadrupled.

Sound travels through solids faster than through air. —This may be nicely illustrated by placing the ear close to the horizontal bar at one end of an iron fence, and having a person at the other end strike the fence a smart blow. Two successive sounds will reach the ear—one through the metal, and afterward another through the air. In some experiments made by Biot, when a bell was struck at the end of an iron tube 3,120 feet long, $2\frac{1}{2}$ seconds elapsed between the two sounds. This would make the velocity in iron nearly ten times that in air.

Sounds travel with the same velocity.—Under ordinary circumstances we see that this must be true. A band is playing at a distance, yet the harmony of the different instruments is preserved. The soft and loud, high and low notes reach the ear at the same time. It has been said that the "heaviest thunder travels no faster than the softest whisper." This is not verified by careful investigations. Mr. Mallet found that in blasting with a charge of 2,000 lbs., the velocity was 967 feet per second, while with a charge of 12,000 lbs. it was increased to 1,210 feet per second. Capt. Parry in his Arctic travels made a similar observation. He states that on a certain occasion when at a considerable distance, the sound of the sunset-gun reached his ears before the officer's word of command to fire, proving that the report of the cannon travelled sensibly faster than the sound of the voice.

The velocity of sound may be used to determine distances.—Light travels instantaneously as far as all distances on the earth are concerned. Sound moves more slowly. For this reason we see a chopper strike with his axe, and a moment elapses before we hear the blow. If the time that intervenes is one second, we know that the distance is about 1,090 feet. By means of the second-hand of a watch or the beating of our pulse, we can count the seconds that elapse between a flash of lightning and the peal of thunder which follows. Multiplying the velocity of sound by the number of seconds, we have the distance of the thunderbolt.

THE INTENSITY OF SOUND *depends upon the amplitude of the vibrations.*—The amplitude is the distance the molecules swing to and fro. As in a pendulum, the greater the amplitude, the greater the velocity. In momentum, we found that the force of a striking body depends upon its weight and its velocity. Just so, if one sound appears to us louder than another, it is because the air molecules hit the ear-drum with greater force. On the top of a mountain, because of the rareness of the atmosphere, there are fewer molecules to strike the ear; hence, according to the principles of momentum, the blow will be less intense.

The intensity of sound diminishes as the square of the distance increases.*—This is the natural effect of the expanding of the sound-wave, which proceeds in the form of a sphere. The larger the sphere, the greater the number of air particles to be set in motion, and hence the feebler their vibration. The surfaces of spheres are proportional to the squares of their radii. The radii of sound-spheres are their distances from the centre of disturbance. Hence the force with which the molecules will strike our ear decreases as the square of our distance from the sounding body.

Speaking-tubes conduct sound to distant rooms because they prevent the waves from expanding and losing their intensity.* Biot found that a conversation in a low tone could be kept up through a Paris water-pipe 3,120 feet long. He says that "it was so

---

* The same proportion obtains in Gravitation, Sound, Light, and Heat. We have seen how the pendulum is based upon the force of gravity, and reveals the laws of falling bodies. Now we find that the pendulum, and even the principles of reflected motion and momentum, are linked with the phenomena of sound. As we progress further, we shall find how Nature is thus interwoven everywhere with proofs of a common plan and a common Author.

easy to be heard, that the only way not to be heard was not to speak at all." A communication could be made in this manner even between two villages. The *ear-trumpet* acts by collecting waves of sound and reflecting them into the ear. The *speaking-trumpet* is often explained on the same principle as the speaking-tube. A more rational theory is, that the sound of the voice is strengthened by the vibrations of the air in the tube.

REFRACTION OF SOUND.—When a sound-wave goes obliquely from one medium to another, it is bent out of its direct course. It may even, like light, be passed through a lens and brought to a focus. B is a thin rubber balloon, filled with carbonic acid gas; w is a watch, and f' is a glass funnel which as-

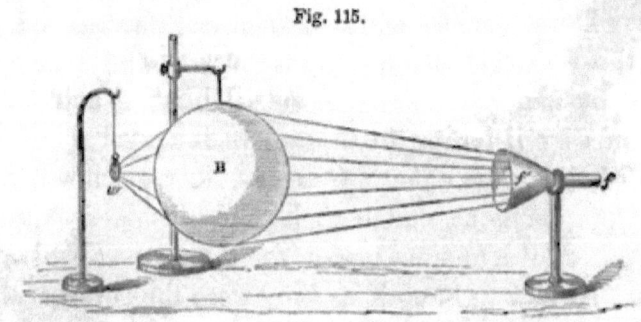

Fig. 115.

sists in collecting the wave at f, where the ear is placed. By moving the head, a point will soon be found where the ticks of the watch can be heard very distinctly, while outside of it they are inaudible.

REFLECTION OF SOUND.—When a sound-wave strikes against the surface of another medium, a portion goes on while the rest is reflected.

The law which governs Reflected Sound is that of Reflected Motion;—*the angle of incidence is equal to that of reflection.* Tyndall relates that a bell on a distant eminence in Heligoland failed to be heard in the town. A reflector was therefore placed behind it, so as to throw the sound-waves in the direction of the long sloping street. This caused every stroke to be distinctly audible. Domes and curved walls and ceilings act in the same manner as mirrors in the reflection of sound. Sir John Herschel relates an amusing illustration of this fact. A confessional in a cathedral in Sicily was so situated that the whispers of the penitents were reflected by the curved roof and brought to a focus at a distant part of the edifice. This point was accidentally discovered by a gentleman, who amused himself and his friends by listening to utterances intended for the ear of the priest alone. One day, however, his wife was the penitent, and both he and his friends were thus made acquainted with family secrets which were as new to himself as they were the reverse of amusing. "The Ear of Dionysius" was a dungeon in Syracuse, so constructed as to convey to the ears of that tyrant every word spoken by its unfortunate inmates. Whispering galleries are commonly made of an elliptical form. Two persons, standing at the foci with their backs to each other, can thus carry on a conversation in whispers which are entirely unnoticed by those between them. Sound-waves have been brought to a focus by the mainsail of a vessel having accidentally taken a concave form; in this manner a bell was once heard 100 miles out at sea.

Decrease of Sound by Reflection.—If we strike the bell represented in Fig. 114, we find a great difference between its sound under the glass receiver and in the open air. Floors are deadened with tan-bark or other fine material; since, as the sound-wave passes from each particle to the next of the unhomogeneous mass, it becomes weakened by partial reflection. During a thunder-storm the air is of such varying density that thunder-peals are never heard at a distance corresponding to their violence. For the same reason, the roar of cannon on a field of battle is not noticeable, and the day has often been lost within a short distance of the reserves of the defeated army, which were waiting for the sound of artillery to call them to the scene of action. The air at night is more homogeneous, and hence sounds are heard more clearly and farther than in the daytime. In foggy weather sounds suffer innumerable reflections from the mist, and are soon destroyed. (*Tyndall.*)

Resonance.—If the reflecting surface be very near, the reflected sound will join the direct one and strengthen it. This effect is termed a resonance. It accounts for the well-known fact that a speaker can be heard much more easily in a close room than in the open air. A smooth wall back of the stand re-enforces the voice in the same manner. The old-fashioned sounding-boards were by no means inefficient, however inelegant may have been their appearance. Shells, by their peculiar convolutions, reflect and augment the various sounds which fill even the stillest air. As we hold them to our ear, they are poetically said to "repeat the murmurs of their ocean-home." Furniture and wall-hangings break up the resonance of a room; and thus our footsteps in unfurnished dwellings sound startlingly distinct. *Echoes* are produced where the reflecting surface is so distant that we can distinguish the reflected sound from the direct one. If the sound be short and quick, this requires at least 56 feet; but if it be an articulate one, 112 feet are necessary. No one can pronounce or hear distinctly more than five syllables in a second; 1,120 ft.$\div$5 = 224 ft.* If the wave travel 224 feet in going and returning, the two sounds will not blend, and the ear can detect a distinct interval between them. A person speaking in a loud voice in front of a mirror, 112 feet distant, can distinguish the echo of the last syllable he utters; if twice that, or 224 feet, the last two syllables, etc. Places where good echoes may be heard abound in every locality. When several parallel surfaces are properly situated, the echo may be repeated back-

---

* This calculation supposes the sound to travel at the rate of 1,120 feet per second.

ward and forward in a surprising manner. At Woodstock, England, is one which repeats 17 syllables by day and 20 by night. The reflecting surface is distant about 2,300 feet; a quick, sharp *ha!* will come back a ringing *ha, ha, ha!* The echo is often softened, as in the Alpine regions, where it warbles a beautiful accompaniment to the shepherd's horn.

THE DIFFERENCE BETWEEN NOISE AND MUSIC is only that between irregular and regular vibrations. Whatever may be the cause which sets the air in motion, if the vibrations be uniform and rapid enough, the sound is musical. If the ticks of a watch could be made with sufficient rapidity, they would lose their individuality and blend into a musical tone. "The puffs of a locomotive are slow on first starting, but they soon increase so as to be almost incapable of being counted. If the puffs could reach 50 or 60 a second, the approach of an engine would be heralded by an organ-peal of tremendous power."

Nothing can be imagined to be more purely a noise than the rattling of a cab over a stony street. The pavement of London is composed of granite blocks, four inches in width. A cab-wheel jolting over this at the rate of eight miles per hour produces a succession of 35 distinct sounds per second. These link themselves together into a soft, deep musical tone, that will bear comparison with notes derived from more sentimental sources. (*Houghton.*)

PITCH.—If we hold a card against the cogs of the wheel in the apparatus shown in Fig. 32, when turned rapidly we shall obtain a pure, clear tone; and the faster the wheel is revolved, the shriller the tone, or the higher the pitch. Hence we conclude that *Pitch depends on the rapidity of the vibrations.*

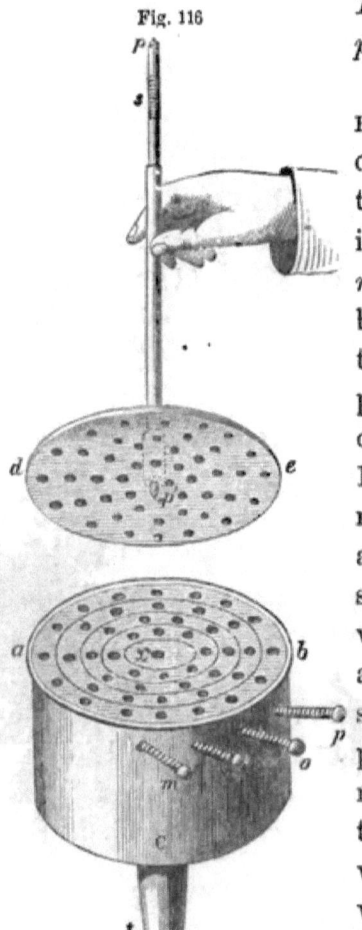

Fig. 116

HOW TO FIND THE NUMBER OF WAVES IN A MUSICAL SOUND.—This is determined by means of an instrument called the *Siren*. C is a cylindrical box; t, a pipe for admitting the air; $a\,b$, a plate pierced with four series of holes, containing 8, 10, 12, and 16 orifices respectively; m, n, o, p are stops for closing any series at pleasure. The vertical rod p is bevelled at p' so as to turn in the socket x; $d\,e$ is a disk pierced with holes corresponding to those in the lower plate, over which it is made to revolve. At s is an endless screw, which, as the axis p revolves, causes two wheels to rotate, and thus turns the hands upon the dial (Fig. 118). On this we can see at any moment the number of revolutions made by the upper disk. The holes in *a b* and *d e* are inclined to each other, so that, when a current of air is forced in at *t*, it passes up through the openings in the lower disk, and striking against the sides of those in the upper disk, causes it to revolve. As the upper disk turns round, it alternately opens and closes the orifices in the lower disk, and thus converts the steady stream of air into uniform puffs. At first they succeed each other so slowly that they may be easily counted. At last, as the motion increases, they link themselves together, and burst into a full, melodious note. As the velocity augments, the pitch rises, until the music becomes so shrill as to be painful. Diminish the speed, and the pitch falls immediately.

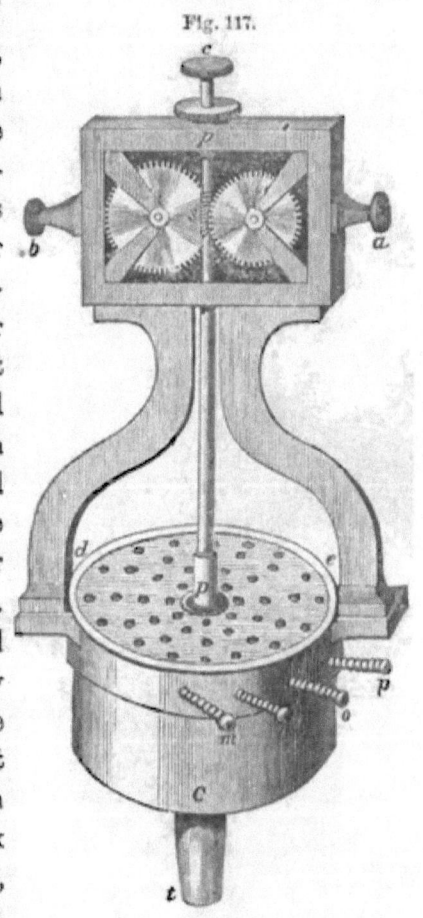

Fig. 117.

Let us now see how the Siren is used to determine the number of vibrations in any sound. Force the air through it steadily until the tone is brought to any required pitch. Find on the dial, at the end of a minute, the number of revolutions made by the disk. When the row containing ten holes is open, and the tone C_2, it will indicate 1,536. There were ten puffs of air, or ten waves of sound, in each revolution. $1,536 \times 10 = 15,360$. Dividing this by 60, we have 256 as the number per second. When the inner and outer rows of holes are opened, the ear immediately detects the difference of an octave between the two sounds. The one containing 8 produces the lower, and the one containing 16 the higher tone. Hence we conclude that an *octave* of any tone is caused by double the number of vibrations.

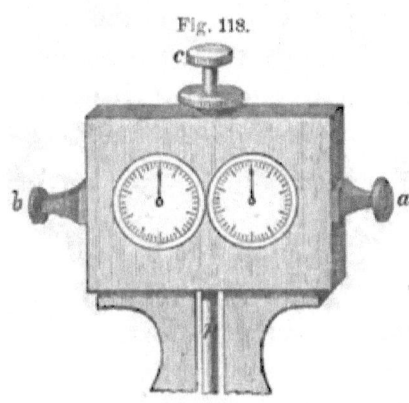

Fig. 118.

How to find the length of the wave in a musical sound.—Suppose the air in the last experiment was of such a temperature that the foremost sound-wave would have reached the distance of 1,120 feet in a second. In that space there were 256 sound-waves. Dividing 1,120 by 256, we have 4 feet 4 inches as the length of each. We see from this that we find the wave-length by dividing the velocity of sound by the number of vibrations per second. As the pitch is elevated by rapidity of vibration, we readily perceive that the low tones in music are produced by the long waves and the high tones by the short waves. An experiment illustrative of this can be made when an express-train passes a railroad station. As the engine approaches us, the waves from the whistle are shortened by the rapid motion, and as it recedes, are lengthened; the pitch of the whistle will therefore be raised as the train comes in, and be lowered as it goes out. The same result may be detected if a person in a high swing produces, while in swift motion, a continuous musical tone upon some instrument.

Application to any Musical Sound.—Whenever notes from any two sources are in unison, they are produced by the same number of vibrations. If the string of a violin, the cord of a guitar, the parchment of a drum, and the pipe of an organ, produce the same musical tone, it is because the vibrations in all are isochronous. "If a voice and a piano execute the same music, the steel strings of the piano and the vocal cords of the singer vibrate together and send out sound-waves of the same length." In order, then, to determine the number and length of the sound-waves produced by a sonorous body, we have only to bring its sound and that of the *siren* in unison. In this way it has been found that the wings of a gnat flap, in flying, at the rate of 15,000 times per second. The waves produced by a man's voice in ordinary conversation are from eight to twelve feet in length, and by a woman's voice from two to four feet. (*Tyndall.*)

SUPER-POSITION OF SOUND-WAVES.—(See Wave Motion, $p. 128$.)—The air may transmit sound-waves from a thousand instruments at once. If the condensation of one wave meet the condensation of another, it will augment the sound, the condensations becoming more condensed and the rarefactions more rarefied by their coincidence. If, on the other hand, the condensation of one meet the rarefaction of the other, the result will be changed; one wave motion will be striving to push the air molecules forward, and the other to urge them backward. Thus, if they meet in exactly opposite phases and the two forces are equal, they will balance each other and silence will ensue. Thus a *sound added to a sound may produce silence.* In the same way, *two motions may produce rest; two lights may cause darkness; and two heats may produce cold.*

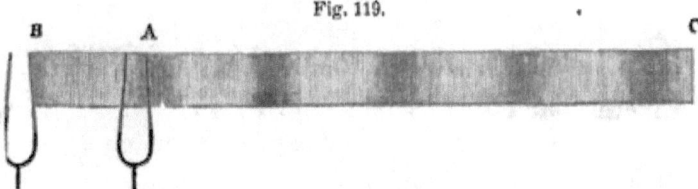

Fig. 119.

Suppose we have two tuning-forks, A and B, placed a wave-length apart, and vibrating in unison.

The waves from the two will coincide, as represented by the light and dark shades in the figure. The same would occur if they were placed at any number of wave-lengths apart. If they are a half wavelength apart, the condensation of A coincides with the rarefaction of B, and *vice versa.* The effect is represented by the uniformity of the shading in Fig. 120. This is termed *interference of sound-waves.*

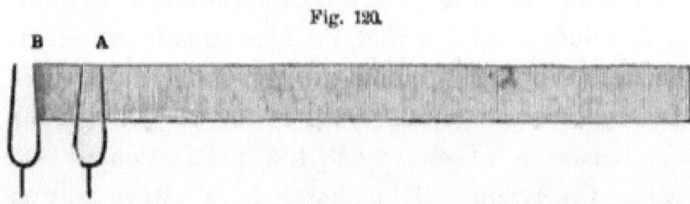
Fig. 120.

There are positions in which the prongs of a tuning-fork interfere with each other so as to produce silence. If we strike the fork and turn it slowly around before the ear, we shall find four points where the interference of the sound-waves entirely neutralizes the vibrations.

VIBRATIONS OF CORDS.—Let $a\ b$ be a stretched

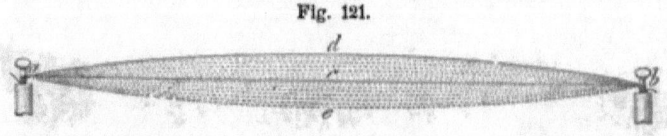
Fig. 121.

cord made to vibrate. The motion from e to d and back again is termed a complete vibration; that from e to d alone, is a half-vibration. The *intensity* of the sound depends on the width of $e\ d$ i. e., the

8

amplitude of the vibration. It is, however, very weak, on account of the small amount of air a simple cord can set in motion. The laws which govern the number of vibrations, and hence the *pitch*, are investigated by means of an instrument known as the SONOMETER. It consists of two cords stretched by weights at P, across two fixed bridges, A and B.

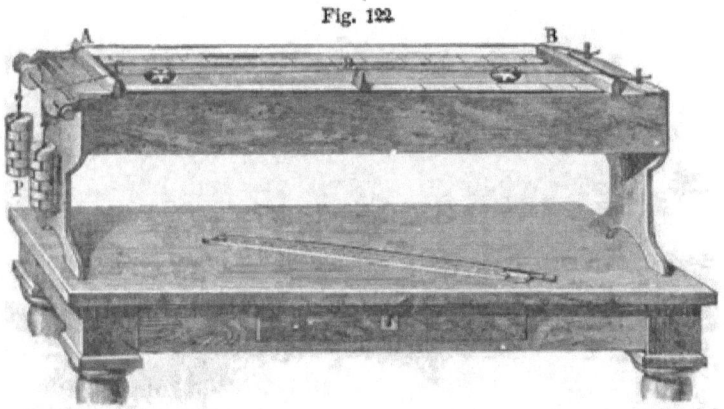

Fig. 122.

D is a movable bridge, which serves to lengthen or shorten the cords at pleasure. Beneath is a wooden box which receives the vibrations of the cords and communicates them to the air within. This is the real sounding body.

1st Law. *The number of vibrations per second increases as the length of the cord decreases.*—Let the cord be caused to vibrate, and we shall hear the note of the entire string. Now place the movable bridge D at the centre of the cord, and we shall obtain a sound the *octave* of the former. Thus by taking one-half the length of the cord we **double the number of vibrations.** If an entire cord makes 20 vibrations per second, one-half will make 40, and one-third, 60. The violin or guitar player elevates the pitch **of any string** of his instrument by moving his finger, and thus shortening the length of the vibrating portion. In the piano, harp, etc., the long and short strings produce the low and high notes respectively.

2d **Law.** *The number of vibrations per second increases as the square-root of the tension.*—The cord when stretched by 1 lb. gives a certain tone: **to double the number of vibrations and obtain the octave requires a weight of 4 lbs.** All stringed instruments are provided with keys, by means of which the tension of the cord and the corresponding pitch may be increased or diminished.

3d Law. *The number of vibrations per second decreases as the square-root of the weight of the cord increases.*—If two strings of the same material be equally stretched, and one have four times the weight **of the other, it will** only vibrate half as often. In the violin the bass notes are produced by the thick strings. In the piano the result is obtained by coiling fine wire around the heavy strings.

NODES.—In the experiments just named, the cord was shortened by means of a movable bridge which held it firmly at the centre. If, instead, we simply rest **the** feather-end of a goose-quill lightly on the string, and then draw the bow over one-half, it will vibrate in two portions and will give the octave as before. Remove the feather, and it will continue to

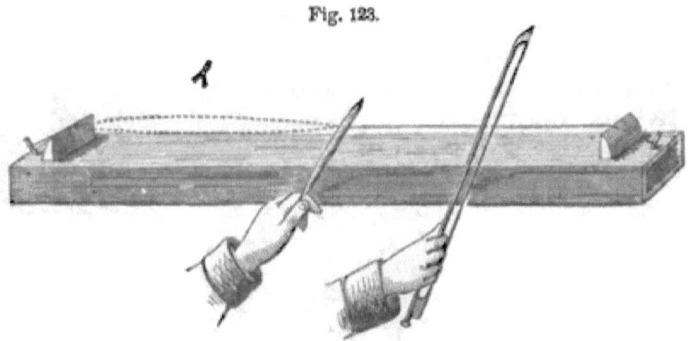
Fig. 123.

vibrate in two parts and to yield the same tone. We can show that the second half vibrates by simply placing across the middle of that portion of the wire a little paper rider. On drawing the bow the

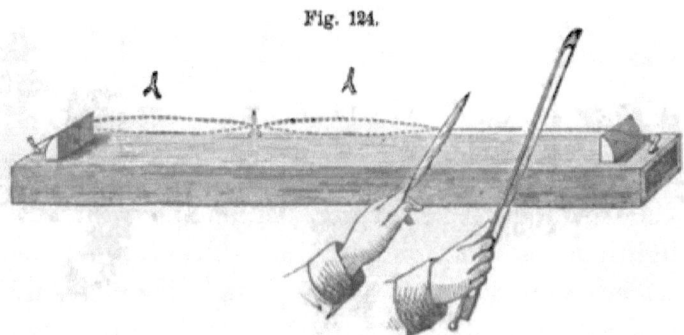
Fig. 124.

rider will be thrown off. Hold the feather so as to separate one-third of the string and cause it to vibrate. The remainder of the cord will vibrate in two segments. When the feather is removed, the entire cord will vibrate in three different parts of equal length, separated by stationary points called *nodes*. This may be shown by placing on the wire three riders; the one at the node will remain, while the others will be thrown off. In the same manner the cord may be divided into any number of equal vibrating segments and stationary nodes.

Acoustic Figures.—Sprinkle some fine sand on a glass or metal plate. Place the finger-nail on one

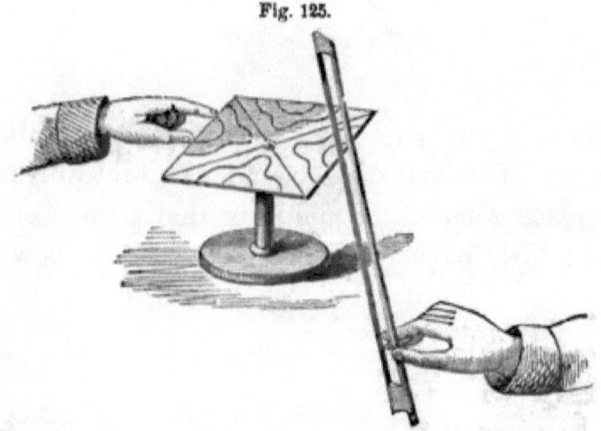

Fig. 125.

edge to stop the vibration at that point, as the feather did in the last experiment, and draw the bow lightly across the opposite edge. The sand will be tossed away from the various parts of the plate and will collect along two *nodal lines*, which divide the large square. It is wonderful to see how the sand will seemingly start into life and dance into line at the touch of the bow. Fig. 126 shows some of the beautiful patterns obtained by Chladni.

174 NATURAL PHILOSOPHY.

Fig. 126.

Harmonics or *overtones.*—Even when a cord is caused to vibrate in its full length, it separates into parts at the same time. Thus we have the full or *fundamental* note of the entire string; and superposed upon that, the higher notes produced by the vibrating parts. These are called *overtones* or *harmonics*. The overtones vary in different instruments. The mingling of the two classes of vibrations determines the *quality* of the sound, and enables us to distinguish the music of different instruments.

Nodes of a Bell.—Let the heavy circle in Fig. 127, represent the circumference of a bell when at rest. Let the hammer strike at $a, b, c,$ or d. At one moment as the bell vibrates it will form an oval with $a\ b$, at the next with $c\ d$ for its longest diameter. When it strikes its deepest note, the bell vibrates in four segments, with n, n, n, n, as the nodal points, whence nodal lines run up from the edge to the crown of the bell. It tends, however, to divide into a greater number of segments, especially if it is very thin, and so to produce a series of harmonic sounds. These overtones, which follow the deep tones of the bell, are frequently very striking, even in a common call-bell.

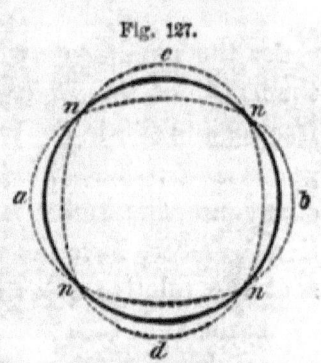

Fig. 127.

Nodes of a Sounding-board.—The case of the violin

or guitar is composed of thin wooden plates which divide into vibrating segments, separated by nodal lines according to the pitch of the note which is being played. The enclosed air vibrating in unison with these, re-enforces the sound and gives it fulness and richness. The sounding-board of the piano acts in a similar manner.

THE MUSICAL SCALE.—The tone produced by the vibrations of an entire string is called its *fundamental* sound. The various sounds of the scale above this are given by the parts of the string indicated by the fractions

$$\begin{array}{cccccccc} C, & D, & E, & F, & G, & A, & B, & C. \\ 1 & 8/9 & 4/5 & 3/4 & 2/3 & 3/5 & 8/15 & 1/2 \end{array}$$

As the number of vibrations varies inversely as the length of the cord, we need but to invert these fractions to obtain the relative number of vibrations per second; thus, $\frac{9}{8}, \frac{5}{4}, \frac{4}{3}, \frac{3}{2}, \frac{5}{3}, \frac{15}{8}, 2$. Reduced to a common denominator, their numerators are proportional, and we have the whole-numbers which represent the relative rates of vibration of the notes of the scale, viz.:

$$24, \ 27, \ 30, \ 32, \ 36, \ 40, \ 45, \ 48.$$

The number of vibrations corresponding to the different letters is,

$$\begin{array}{cccccccc} C, & D, & E, & F, & G, & A, & B, & C. \\ 128, & 144, & 160, & 170, & 192, & 214, & 240, & 256. \end{array}$$

WIND INSTRUMENTS produce musical sounds by means of enclosed columns of air. Sound-waves run backward and forward through the tube and act on the surrounding air like the vibrations of a cord. The sound-waves in organ-pipes are set in motion either by means of fixed mouth-pieces or vibrating reeds. The air is forced from the bellows into the

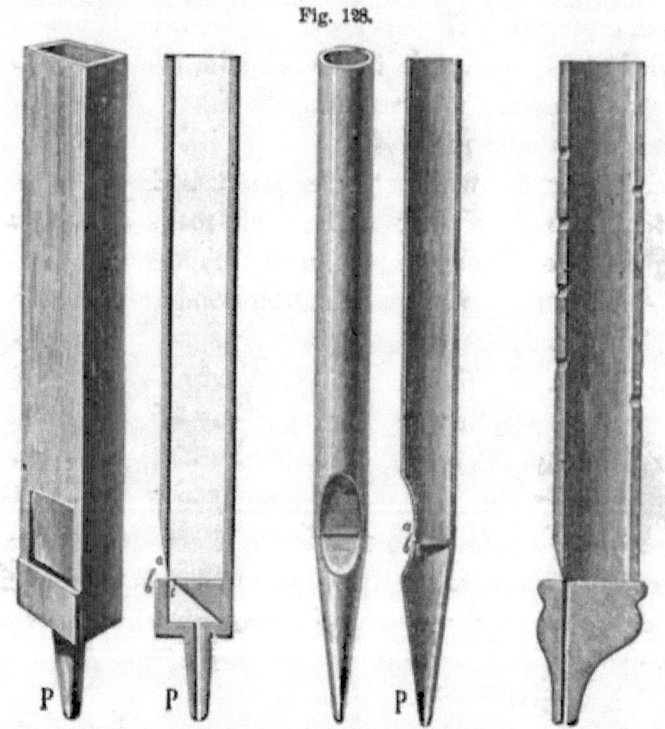

Fig. 128.

tube P, through the vent i, and striking against the thin edge a, produces a flutter. The column of air above, being thus thrown into vibration, re-enforces the sound and gives a full musical tone. The length of the pipe determines the pitch. The variation in the quality of different wind-instruments is caused by the mingling of the harmonics with the fundamental tone. In the flute, for example, the vibrating column of air may be made to break up into vibrating segments with stationary nodes, by merely varying the force of the breath.

SYMPATHETIC VIBRATIONS.—Stand near a piano and produce a musical tone with the voice, and you will find that a certain wire selects that pulse of sound and responds to it. Change the pitch, and the first string ceases, while another replies. If a hundred tuning-forks of different tones be made to sound at the foot of an organ-pipe, it will choose the one to which it is able to reply, and respond to that alone. Two clocks set on one shelf or against the same wall, affect each other. Watches in the shop-window keep better time than when carried singly.

THE EAR.—Fig. 129 is a sketch of the ear, drawn from a model. The small bones are much magnified, in order to give a distinct idea of their shape. A F represents a rear view of the outer ear. The sound-wave passes into the auditory canal, which is about one inch in length, and striking against the *tympanum* or drum, E, which closes the orifice

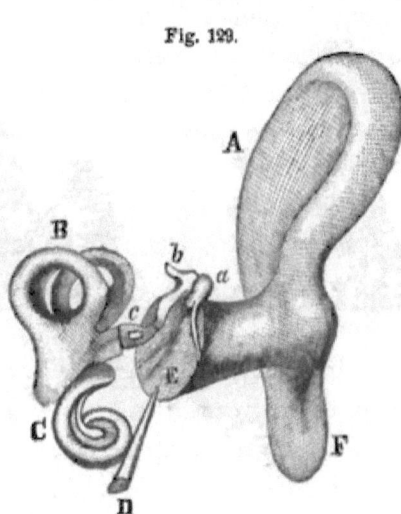
Fig. 129.

of the external ear, throws this membrane into vibration. Next, the series of small bones, *a, b, c,* called respectively, from their peculiar form, the *hammer, anvil,* and *stirrup,* conduct it to the inner ear, which is termed, from its complicated structure, the *labyrinth.* This is filled with liquid, and contains the semi-circular canals, B, and the cochlea (*snail-shell*), C, which receive the vibrations and transmit them to the auditory nerve, the fine filaments of which are spread out to catch every pulsation of the sound-wave. The middle ear, which contains the chain of small bones, is a simple cavity about $\frac{1}{2}$ inch in diameter, filled with air. It communicates with the mouth by means of the Eustachian tube, D. Within the labyrinth are also fine, elastic hair-bristles and crystalline particles among the nerve-fibres, wonderfully fitted, the one to receive and the other to prolong the vibrations; and lastly, a *lute* of 3,000 microscopic strings, so stretched as to vibrate in unison with any sound. The Eustachian tube is generally closed, thus cutting off the air in the inner cavity from the external air. If at any time the pressure of the atmosphere without becomes greater or less than that within, the tympanum feels the strain, pain is experienced, and partial deafness ensues. A forcible concussion frequently produces in this way a temporary deafness. In the act of swallowing, the tube is opened and the equilibrium restored. We may force air into the cavity of the ear by closing our mouth and nose, and forcibly expiring the air from our lungs. This will render us insensible to low sounds, as the rumble of a railway-train, while we can hear the higher ones as usual.

LIMITS OF HEARING.— Helmholtz fixes the lowest limit of musical sounds at 16 vibrations per second,* and the highest at 38,000. Below this number the pulses cease to link themselves together, and become distinct sounds. The range of the ear is thus about eleven octaves. The practical range of music is, however, only about seven octaves. The capacity to hear the higher tones varies in different persons. A sound which is entirely audible to one may be utter silence to another. Some ears cannot distinguish the squeak of a bat or the chirp of a cricket, while others are acutely sensitive to these shrill sounds. Indeed, the auditory nerve seems generally more alive to the short quick vibrations than to the long slow ones. The whirr of a locust is much more noticeable than the sighing of the wind through the trees. To this, however, there are noticeable exceptions. The author knows of a person who is entirely insensible to the higher tones of the voice, but acutely sensitive to all the lower ones. Thus on one occasion, being in a distant room, she did not notice the ringing of the bell announcing dinner, but heard the noise the bell made when returned

---

* A tone produced by about 32 vibrations per second may be made by inserting the finger lightly into the ear, bringing at the same time the muscles of the hand into strong contraction. A sound will be heard which is as deep as the toll of a cathedral bell.

to its place on the shelf. A continuous blast of air has no effect to produce sound. The rush of the grand aërial rivers above us we never hear. They flow on ceaselessly but silently in the upper regions of the air. A whirlwind is noiseless. Let, however, the great billows strike a tree and wrench it from the ground, and we can hear the secondary shorter waves which set out from the struggling limbs and the tossing leaves.

Our unconsciousness is no proof of the absence of sound. There are, doubtless, sounds in Nature of which we have no conception. Could our sense be quickened, what celestial harmony might thrill us! Professor Cooke beautifully says: "The very air around us may be resounding with the hallelujahs of the heavenly host, while our dull ears hear nothing but the feeble accents of our broken prayers."

The ability of the ear to detect and analyze sound is wonderful beyond all comprehension. Sound-waves chase each other up and down through the air, superposed in entangled pulsations, yet a cylinder of the air not larger than a quill conveys them to the ear, and each string of that wonderful harp selects its appropriate sound, and repeats the music to the soul within. Though a thousand instruments be played at once, there is no confusion, but each is heard, and all blend in harmony.

THE TENDENCY OF NATURE TO MUSIC.—" Friction," says Tyndall, " is rhythmic." A bullet flying through the air sings softly as a bird. The limbs and leaves of trees murmur as they sway in the breeze. The rumble of a great city, all the confused noises of nature when softened by distance, are said to be on one pitch—the key of F. Falling water, singing birds, sighing winds, everywhere attest that the same Divine love of the beautiful which causes the rivers to wind through the landscape, the trees to bend in a graceful curve—the line of beauty—and the rarest flowers to bud and blossom where no eye save His may see them, delights also in the anthem of praise which Nature sings for His ear alone.

SENSITIVE FLAMES.—Flames are frequently extremely sensitive to certain sounds. At an instrumental concert the gaslights often vibrate in unison with certain pulsations of the sound which they seem to select. This is most noticeable when the pressure of gas is so great that the flame is just on the verge of flaring, and the vibration of the soundwave is sufficient, as it were, to "push it over the precipice." If we turn on the gas, in a common fish-tail burner, we reach a point where a shrill whistle will produce the same effect as increased pressure of the gas, and cause the jet to thrust out long, quivering flames. Prof. Barret, of London, describes a peculiar jet which was so "sensitive that it would tremble and cower at a hiss, like a human being, and even beat time to the ticking of a watch."

SINGING FLAMES.—If we lower a glass tube over a

small jet of gas, we soon reach a point where the flame leaps spontaneously into song. At first the sound seems far remote, but gradually approaches until it bursts into a shrill scream that is almost intolerable. The length of the tube and the size of the jet determine the pitch of the note. If we raise the tube to a point where it is just ready to sing, we shall find that it will respond to the voice when the proper note is struck.*

The flame, owing to the friction at the mouth of the pipe, is thrown into vibration. The air in the tube, being heated, rises, and not only vibrates in unison with the jet, but, like the organ-pipe, selects the tone which is adapted to its length, and in part governs the pulsations of the flame.

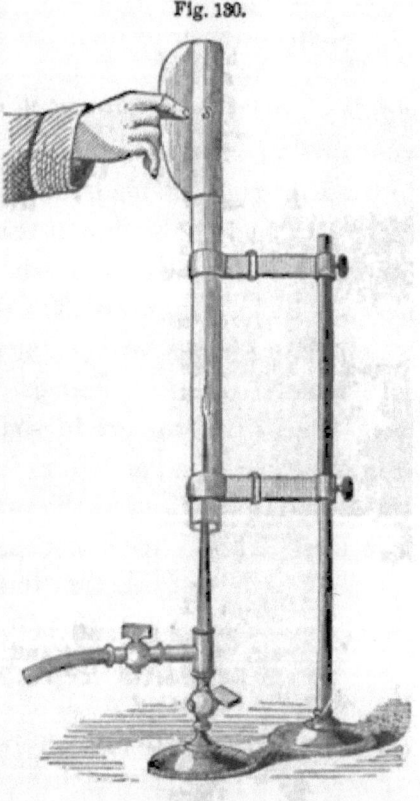

Fig. 130.

---

* See Rev. Chem., p. 55. The jets are made by drawing out glass tubing to a fine point over a spirit-lamp. The length of the tube may be varied, as in the figure, by means of a paper tube.

Practical Questions.—1. Why cannot the rear of a long column of soldiers keep time to the music in front? 2. Three minutes elapse between the flash and report of a thunderbolt; how far distant is it? 3. Five seconds expire between the flash and report of a gun; what is its distance? 4. Suppose a speaking-tube should connect two villages ten miles apart; how long would it take the sound to travel that distance? 5. The report of a pistol-shot was returned to the ear from the face of a cliff in four seconds; what was the distance? 6. What is the cause of the difference between the voice of man and woman? A base and a tenor voice? 7. What is the number of vibrations per second necessary to produce the fifth tone of the scale of C? 8. What is the length of each sound-wave in that tone when the temperature is at zero? 9. What is the number of vibrations in the fourth tone above middle C? 10. A meteor of Nov. 13, 1868, is said to have exploded at a height of 60 miles; what time would it have required for its sound to reach the earth? 11. A stone was let fall into a well, and in four seconds was heard to strike the bottom; how deep was the well? 12. What time will it require for a sound to travel five miles in the still water of a lake? 13. How much louder will be the report of a gun to an observer at a distance of 20 rods than to one at half a mile? 14. Does sound travel faster at the foot or at the top of a mountain? 15. Why is an echo weaker than the original sound? 16. Why is it so fatiguing to talk through a speaking-trumpet?

Curious Facts in Sound.—Silliman says the roar of cannon has been heard at a distance of 250 miles by putting the ear to the ground. In Capt. Parry's third Expedition, Lieut. Forster carried on a conversation with a man at a distance of 1¼ miles. The sentinel's "All's well" has been heard from Old to New Gibraltar, a distance of 10 miles. The cannonading at the battle of Jena was heard at Dresden, 92 miles away. The celebrated echo of the Metelli at Rome was capable of repeating the first line of the Æneid 8 times distinctly In Fairfax County, Va., is an echo which will return 20 notes played on a flute, but supplies the place of some notes with their thirds, fifths, or octaves. Sir John Herschel says the tick of a watch may be heard from one end of the Abbey Church of St. Albans to the other. At Carisbrook Castle, in the Isle of Wight, is a well 210 feet deep and 12 feet wide. It is lined with smooth masonry. When a pin is dropped into the well it is distinctly heard to strike the water. In certain parts of the Colosseum at London the tearing of paper sounds like the patter of hail, while a single exclamation comes back a peal of laughter.

A tired bee hums on E, while in pursuit of honey it hums contentedly on A. The common horse-fly, when held captive, moves its wings 335 times a second; a honey-bee, 190 times.

Youmans says it is marvellous how slight an impulse throws a vast amount of air into motion. We can easily hear the song of a bird 500 feet above us. For its melody to reach us it must have filled with wave-pulsations a sphere of air 1,000 feet in diameter, or set in motion 18 tons of the atmosphere.

On Light.

The sunbeam comes to the earth as simply motion of ether-waves, yet it is the only source of beauty, life, and power. In the growing plant, the burning coal, the flying bird, the glaring lightning, the blooming flower, the rushing engine, the roaring cataract, the pattering rain—we see only varied manifestations of this one all-energizing force.

OPTICS.

Optics treats of Light.

Definitions.—A *luminous* body is one that emits light. A *non-luminous* body is one that reflects light, and is visible only in the presence of a luminous body. A *medium* is any substance through which light passes. A *transparent** body is one that offers so little obstruction to the passage of light that we can see objects through it. A *translucent* body is one that lets some light pass, but not enough to render objects visible through it. An *opaque* body is one that does not transmit light. A *ray of light* is a single line of light; it may be traced in a dark room into which a sunbeam is admitted.

---

* Though we speak of transparent and opaque substances, these terms are merely relative. No body is perfectly transparent, nor is any entirely opaque. Glass obstructs some light. It is said that if the atmosphere were 700 miles deep, no light would reach our eye. Deep-sea dredging has recently shown that light penetrates water to great depths, so that even the Atlantic cable may not lie in an abyss of utter darkness. On the other hand, gold, when beaten into leaf, becomes translucent, and appears of a faint green color; and horn, when scraped becomes semi-transparent.

by the floating particles of dust which reflect the light to the eye.

THE VISUAL ANGLE is the angle formed at the eye by rays coming from the extremities of an object. The angle A O B is the angle of vision subtended

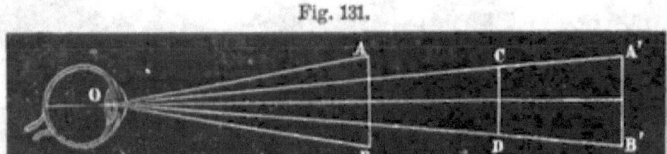

Fig. 131.

by the object A B. The size of this angle varies with the distance of the body. A B and A' B' are of the same length, and yet the angle A' O B' is much smaller than A O B, and hence A' B' will seem much shorter than A B. We estimate the distance and size of objects in various ways, but the two are intimately connected, since we have by long experience learned to associate them. Knowing the distance of an object, we determine its size immediately from the visual angle. We can vary the apparent size of any body at which we are looking, if by any means we increase or diminish this angle. This principle will be found of great importance in the formation of images by mirrors and lenses.

LAWS OF LIGHT.—1. Light passes off from a luminous body equally in every direction.

2. Light travels through a uniform medium in straight lines.

3. The intensity of light decreases as the square of the distance increases.

THE VELOCITY OF LIGHT is about 183,000 miles per second. This is so great that for all distances on the earth it is instantaneous. A sunbeam would girt the globe quicker than we can wink. This rate has been determined in various ways, but a most simple method is thus explained. The planet Jupiter has four moons; as these pass around the planet, they are eclipsed from our sight at regular intervals. In the cut, let J represent Jupiter, *e* one

Fig. 132.

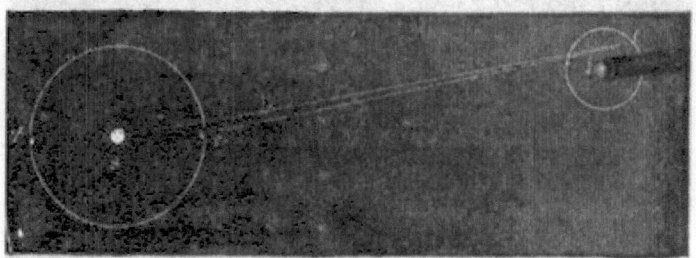

of the moons, S the sun, and T and *t* different positions of the earth as it moves in its orbit around the sun. Römer noticed that when the earth was at T, the eclipse occurred 16 min. and 36 sec. earlier than when at *t*. He could account for this only on the supposition that it requires that time for the light to travel across the earth's orbit. This distance is 183,000,000 miles. Hence the velocity is about 183,000 miles per second.

THE UNDULATORY THEORY OF LIGHT.—There is supposed to be a fluid, termed *ether*, constituting a kind of *universal atmosphere*, diffused throughout all space. It is so subtle that it fills the pores of all bodies, eludes all chemical tests, passes in through the glass receiver and remains even in the vacuum of an air-pump. A luminous body sets in motion waves of ether, which pass off in every direction. These move at the rate of 183,000 miles per second, and breaking upon the eye, give to us the impression of sight. This etherial wave-motion is precisely like that of sound, except that the vibrations are *transverse* (crosswise) to the line of direction. Thus, if we suppose a star directly overhead and a ray of light coming down to us, we should conceive that the particles which compose the waves are vibrating N. S. E. W., and toward every other point of the compass all at once.

REFLECTION OF LIGHT.

DIFFUSED AND REFLECTED LIGHT.—When light falls upon any surface, one portion is transmitted and another is reflected. The law is that of Motion and Sound—*i. e.*, "The angle of incidence is equal to the angle of reflection." When the surface is rough, the multitude of little protuberances scatter the rays in every conceivable direction, and we can therefore see such a body from any point. This forms what we term *diffused* light. When the surface is smoothed and polished, the rays are more uniformly reflected in particular directions, and, when we stand in the proper position, will bring to us the images of other objects. We thus view all non-luminous objects by means of irregularly reflected (diffused) light, and images of objects by means of regularly reflected light. However, the most perfectly polished substance diffuses some light—enough to enable us to trace its surface; were it not so, we could not be aware of its existence. The deception of a mirror is oftentimes very nearly complete; yet a little dust or vapor, increasing the irregular reflection, will at once bring the surface to view.

REFLECTION VARIES WITH THE ANGLE.—We notice this very clearly if we look at the images of objects in still water. Those which are near us are not as distinct as those on the opposite bank, because the rays from the latter strike the water more obliquely than the former, and so are more perfectly reflected to the eye. The image of the sun at mid-day is not so bright as when it is near the horizon.

MIRRORS.—All reflecting surfaces are termed mirrors. These are of three kinds—*plane*, *concave*, and *convex*. The first has a flat surface; the second, one like the inside, and the third, one like the outside of a watch-crystal.

The general principle of mirrors is, that *the image is always seen in the direction of the reflected ray as it enters the eye.*

PLANE MIRRORS.—Rays of light retain their relative direction after reflection from a plane surface.*

---

* The effect of the various mirrors is best understood if we draw a mirror of the kind under consideration, and then repre-

An image seen in a plane mirror is therefore erect and of the same size as the object. It is, however, reversed right and left.

Why the image is seen as far behind the mirror as the object is in front of it.—Let A B be an arrow held in front of the mirror, M N. Rays of light from the point A striking upon the mirror at C, are reflected, and enter the eye as if they came from *a*. Rays from B, in the same manner, seem to come from *b*. Since the image is seen in the direction of the reflected rays, it appears at *a b*, a point as far behind M N as the real arrow is in front of it.

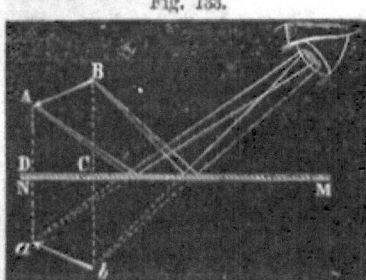

Fig. 133.

Why we can see several images of an object in a mirror.—Metallic mirrors form but a single image. If, however, we look obliquely at the image of a candle

---

sent rays of the different classes, erect perpendiculars at the point of incidence and find the reflected rays. A little practice of this kind will benefit more than any description. It will aid in drawing the perpendicular to a convex or concave surface, if we remember that it is always a radius of the sphere of which the mirror forms a part. A book held in various positions before a common looking-glass should be used to illustrate the action of plane mirrors. Many of the effects of concave and convex mirrors may be seen on the inner and outer surface of a bright spoon, a call-bell, or a metal cup. Much instructive amusement may be obtained in the examination of the curious and grotesque figures thus revealed.

in a looking-glass, we shall often see several images, the first one very feeble, the next bright, and the others gradually diminishing in intensity. The ray from A is in part reflected to the eye from the glass at *b*, and gives rise to the image *a*; the remainder passes on and is reflected from the metallic surface at *c*, and coming to the eye forms a second image *a'*.

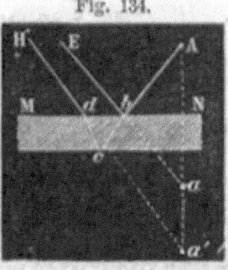

Fig. 134.

The ray *c d*, when leaving the glass at *d*, loses a part, which is reflected back to form a third image. This ray in turn is divided to form a fourth, and so on.

Fig. 135.

Images seen in water are symmetrical, but inverted. The reason of this is best understood by holding an object in front of a horizontal looking-glass and noticing the angle at which the various rays must strike the surface in order to be reflected to the eye. When the sun or moon is shining high in the heavens, we always see the image in the water at only one spot, while the rest of the surface appears dark. The light falls upon all parts, but the rays are reflected at the right angle to reach the eye from one point alone. Each observer sees the image at a different place. When the surface of the water is ruffled, a long tremulous line of light is reflected from the side of each tiny wave that is turned toward us. As each little billow rises, it flashes a gleam of light to our eye, and then sinking, comes up beyond, only to reflect another ray.

A CONCAVE MIRROR *tends to collect the rays of light.*—The point where the reflected rays meet is termed the *principal focus* (*focus*, a hearth). It is half way between the mirror and the centre of curvature—*i. e.*, the centre of the hollow sphere of which the mirror is a part. In Fig. 136 we have parallel rays falling upon a mirror. C is the centre of curvature; F, the principal focus, half-way between A and C; A F,

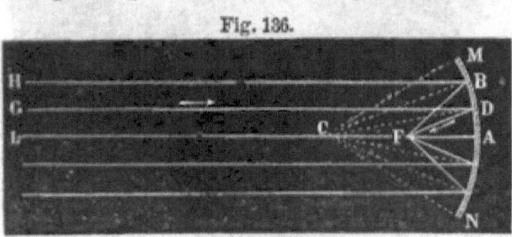

Fig. 136.

the focal distance; C B, C D, etc., radii of the sphere (perpendiculars, to find the angle of incidence); the angles H B C, G D C, etc., equal respectively to F B C,

F D C, etc. A light held at C will have its rays brought to a focus at C; on the other hand, one at F will be reflected in a beam of parallel rays.

Images formed by concave mirrors.—Hang a concave mirror against the wall, and stand closely to it between the mirror and the principal focus. The image is *erect* and much *larger than life.* The ray *a* falls upon the mirror, is reflected and strikes the

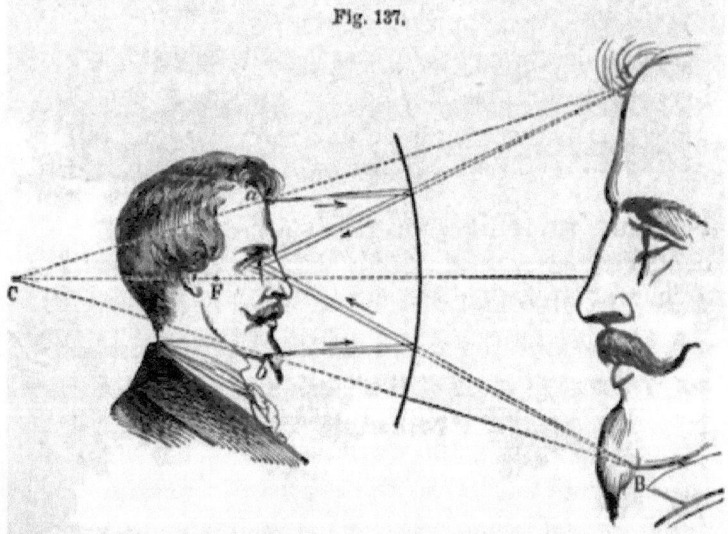

Fig. 137.

eye as if it came from A. In the same manner *b* is seen at B. The visual angle is increased the nearer we approach the mirror, and hence the greater the magnifying power. We now walk back. When we reach the focus, the image becomes blurred, and finally disappears. We are in the position of the candle *a b* (Fig. 138), and the real image is behind us at A B. A few of the parallel rays, however, enter the eye, and an indistinct image is formed. Retiring still farther, we come to the centre of curvature. Here we find no distinct image, although portions of our figure, as we catch snatches of the rays which are forming the image A B, are magnified in the most uncouth and absurd manner. As we

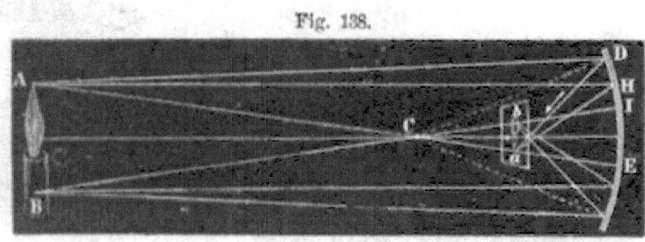

Fig. 138.

continue to recede, we reach a point beyond the centre of curvature. Here we occupy the position A B (Fig. 138), and we see the image in the position *a b*, in front of the mirror inverted. It is inverted since the rays cross at the focus, and is smaller than life because the visual angle is diminished. The positions occupied by the two candles, *a b* and A B, are termed *conjugate foci*, since an object at either point is brought to a focus at the other.

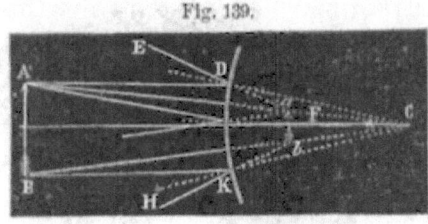

Fig. 139.

A CONVEX MIRROR *tends to scatter the rays of light.*— In the figure we notice how the parallel rays A D and B K are reflected in the diverging lines D E and K H. An eye receiving these rays will perceive the image of A B at *a b*, erect, and smaller than the object since the visual angle is diminished.

TOTAL REFLECTION.—When we look very obliquely into a pond, we cannot see the bottom, because the rays of light from below are reflected downward at the surface. If we look up into a glass of water, we shall see the upper surface gleaming like burnished silver. This effect occurs only when light passes at a definite angle from a denser to a rarer medium. It is termed *total*, since, unlike the other cases of reflection, all the light is bent back.

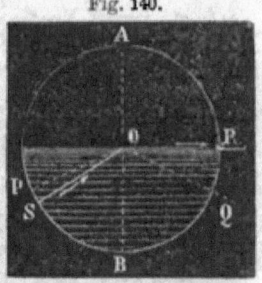

Fig. 140.

REFRACTION OF LIGHT.

We have already seen that when a ray of light passes from one medium to another of different density, one portion is irregularly reflected, and by that means the surface is made visible; that, if the surface is smooth, the larger part is regularly reflected, and in that way images of objects are seen. We now speak of the portion which passes on to the next medium, and which is *refracted* or bent out of its course.

ILLUSTRATIONS OF REFRACTION.—A spoon in clear tea appears bent. An oar dipping in still water seems to break at the point where it enters the water. Fish seem nearer the surface than they really are, and Indians, who spear them, always try to strike perpendicularly, or else aim lower than they apparently lie. Water is always deeper than it appears. Look obliquely and steadily into a pail of water, then place your finger on the outside where the bottom seems to be; you will be surprised, on examination, to find that the real bottom is several inches below your finger. Fill a glass dish with water, and, darkening the windows, let a single sunbeam fall upon the surface. The ray will be seen to bend as it enters. A little chalk-dust scattered through the air will make the beam very distinct. Place a cent at the bottom of a bowl. Standing where you cannot see the coin, let another person pour water into the vessel, when the coin will be apparently lifted into view.

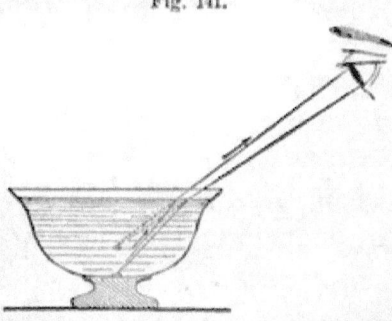

Fig. 141.

Let L, Fig. 142, be a body beneath the water. A ray, L A, coming to the surface, is bent downward toward C, and strikes the eye as if it came from L'. The object will therefore be elevated above its true place. In order to understand the apparent change of position produced by refraction, we have but to remember this principle—*the object is*

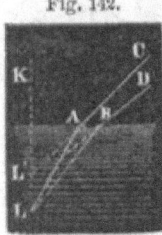

Fig. 142.

always seen in the direction of the refracted ray as it enters the eye.

LAWS OF REFRACTION.—1. In passing into a rarer medium, the ray is bent from the perpendicular. 2. In passing into a denser medium, the ray is bent toward the perpendicular.

PATH OF RAYS THROUGH A WINDOW-GLASS.—When a ray enters a window-glass, it is refracted toward the perpendicular (2d law), and on leaving, is equally refracted from the perpendicular (1st law). The general direction of objects is therefore not changed. A poor quality of glass, however, produces distortion by its unequal density and uneven surface.

Fig. 143.

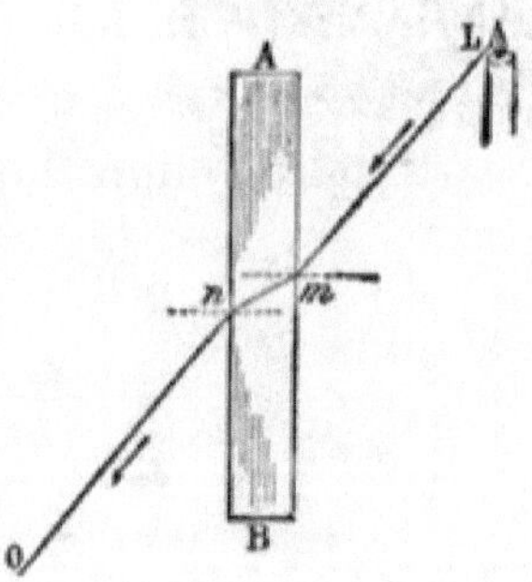

PATH OF RAYS THROUGH A PRISM.—A ray of light, on entering and leaving a prism, is refracted as in the case of a window-glass. The inclination of the sides, however, causes the ray to be *bent twice in the same direction.* The candle L will therefore appear to be at *r*.

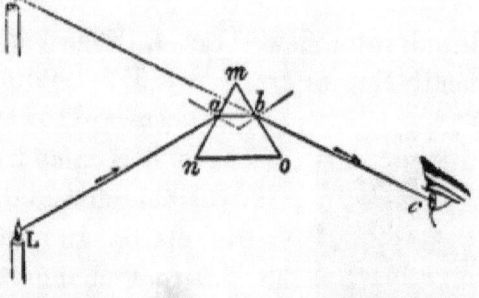

Fig. 144.

LENSES.—A lens is a transparent body, with at least one curved surface. There are two general classes of lenses, *concave* and *convex*. Six varieties of these are used in optical instruments, viz., the

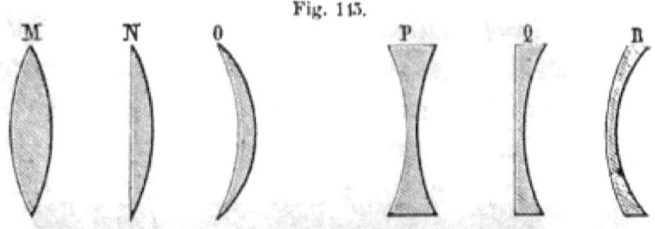

Fig. 145.

double-convex (M), the plano-convex (N), the meniscus (O), the double-concave (P), the plano-concave (Q), and the concavo-convex (R).

THE DOUBLE-CONVEX LENS has two convex surfaces. Rays of light falling upon a *convex* lens, as upon a *concave* mirror, tend to converge to a focus. A ray

Fig. 146.

passing along the line X (the axis of the lens), as it strikes the surface perpendicularly, is not refracted. The parallel rays M, L, etc., are refracted both on entering and on leaving the lens, and are brought together at F, the *focus*. If a light were placed at F, its rays would be refracted in parallel lines.

The convex lens is sometimes termed a *burning-glass*. It is used, like the concave mirror, for collecting the sun's rays, and hence is a ready means of obtaining fire. Lenses have been manufactured of sufficient power to fuse the metals. One, of two feet in diameter, made at Leipsic in 1691, melted plate-iron, and converted a piece of burnt brick into yellow glass. *The image formed by a convex lens* is, in size and position, precisely like that

Fig. 147.

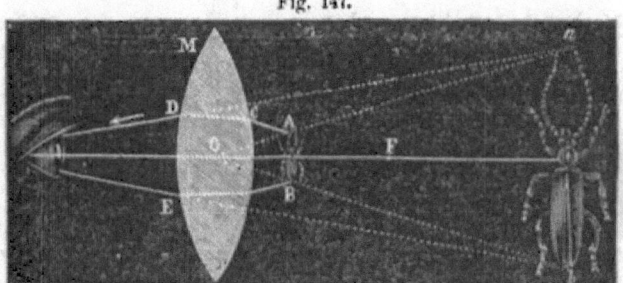

seen in a *concave* mirror. If we hold a lens above a printed page, when we obtain the focal distance correctly, we shall find the letters right-side up and highly *magnified*. By an inspection of the figure we see that the converging power of the lens simply increases the visual angle, and thus makes the object A B appear the size *a b*. Moving the lens back from the page, the letters disappear entirely as we pass the principal focus. At length they suddenly reappear again, but *smaller* and *inverted*. By examining Fig. 148, we see how the rays from A B cross each other at the focus, and thus invert the image *a b*, at the same time reducing the visual angle.

9*

Fig. 148.

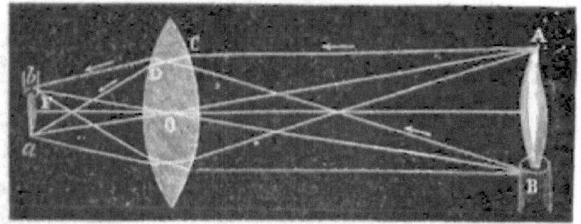

THE DOUBLE-CONCAVE LENS has two concave surfaces. Rays of light falling upon a double *concave* lens, as upon a *convex* mirror, are scattered. Thus,

Fig. 149.

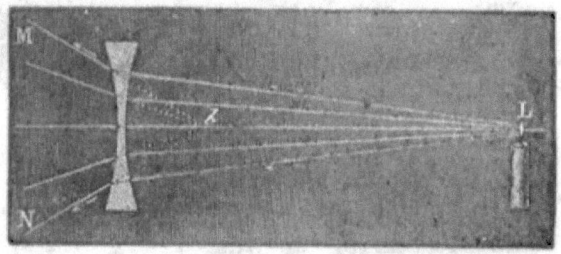

diverging rays from a light at L are rendered more diverging; and, to an eye which receives the rays M N, the candle would be located at l, where the image would be seen.

The image formed by a concave lens is, in size and position, like that seen in a *convex* mirror. The visual angle is decreased, and the rays do not cross; hence the image of A B is seen at $a\ b$, erect, and diminished in size (Fig. 150).

MIRAGE is an optical delusion whereby pictures of distant objects are seen as if near. On the heated deserts of Africa, the traveller beholds quiet lakes and shadows of trees in their cool waters. Rushing

Fig. 150.

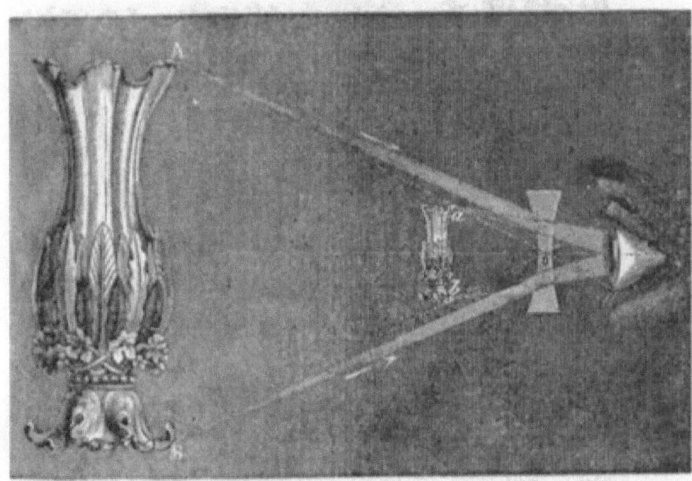

forward to slake his eager thirst, he finds only the barren waste of sand. The mariner often sees in the sky the images of ships, and the far-distant coast, with its familiar cliffs and shipping, so perfect in outline as to be instantly recognized.

The cause is found in the reflection and refraction of the rays of light as they traverse layers of air of unequal density. In Fig. 151, rays of light from a clump of trees at the left are reflected from a layer at a, and enter the eye of the Arab as if they came from the surface of a lake below. The sandy desert itself, shimmering in the hot sun, looks in the distance not unlike the surface of tranquil water. Sometimes, at sea, a layer of air high up in the sky acts as a total reflector, and sends down an inverted image of ships which are far beyond the horizon.

Fig. 151.

THE COMPOSITION OF LIGHT.

SOLAR SPECTRUM.—When a sunbeam is allowed to shine through a prism, the ray is not only bent from its course, as we have already seen, but is spread out, fan-like, into a broad band of rainbow-colors, called the *solar spectrum*. It contains the seven primary colors—Violet, Indigo, Blue, Green, Yellow, Orange, Red. (These may be remembered in their order, by noticing that the initial letters spell the absurdly meaningless word, Vib-gyor.) If we receive the spectrum on a concave mirror, or pass it through a double-convex lens, it will form a white spot. We therefore conclude that white light is composed of seven different colors.*

---

* Many hold, with Sir David Brewster, that there are but three primary colors—*red, yellow*, and *blue*. It is often convenient for purposes of explanation to thus consider it. Helmholtz denies the

They are separated because the prism bends them unequally. The violet is most refracted and the red

Fig. 152

least. The *dispersive power* of a prism, *i. e.*, its ability to spread apart the colored rays, depends on the material of which it is made. A flint-glass prism is

---

truth of this in toto, although he admits that all colors can be produced from the three. On the other hand, John Herschel claims to have discovered an *eighth* color below the red, which is of a crimson hue, and a *ninth* beyond the violet, which is of a lavender hue. Professor Stokes in addition believes in a *tenth* color beyond the lavender, which he styles the fluorescent ray, as it resembles the shade of some kinds of fluor spar.

commonly used. A hollow one, filled with oil of cassia or bisulphide of carbon, has far higher dispersive power.

THREE CLASSES OF RAYS IN THE SOLAR SPECTRUM.—These are the *calorific*, or heat-rays; the *colorific*, or luminous rays; and the *actinic*, or chemical rays. If we examine the spectrum with a delicate thermometer, we shall find that the heat increases gradually from the violet toward the red, and becomes the greatest in the dark space just beyond. If we test with a paper containing nitrate of silver, it will blacken least in the red, more toward the violet, and most in the dark space beyond. Artificial lights differ in the proportion of the three classes of rays. Seeds will sprout best under a blue glass. Red is the warmest color. A photograph could be taken in the dark by means of the chemical rays alone. A soldier dressed in gray or green clothing is less liable to be shot than one in red or yellow.

COMPLEMENTARY COLORS.—Two colors, which by their mixture produce white light, are termed complementary to each other. Let us suppose, for simplicity of statement, that white light is composed of the three colors, red, yellow, and blue. Then, since yellow and blue, when mixed, form green, we have red and green as complementary colors. Red and blue produce violet; hence yellow and violet are complementary colors. If we look steadily at a colored wafer lying on a sheet of white paper, we shall see a fringe of the comple-

OPTICS.   207

mentary color play-
ing about it. If we
watch bright red
clouds, the patches
of clear sky will seem
green. In examining
ribbons of the same
color, the eye be-
comes wearied and
unable to detect the
shade, because of the

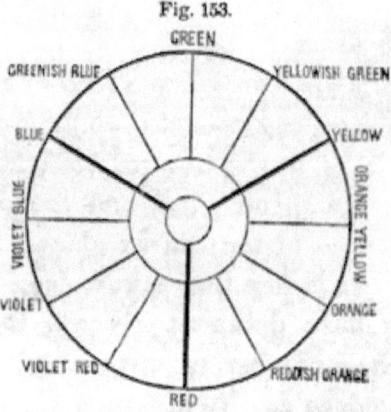

Fig. 153.

mingling of the complementary color. A knowledge
of this subject is very essential to the harmony of
colors in painting, or the arrangement of a bouquet,
that the result may be pleasing to a cultivated taste.

INTERFERENCE OF LIGHT.—*Newton's Rings.*—The
convex side of a plano-convex lens is pressed down
upon a flat surface of glass.
The two surfaces will touch
each other at the centre; and

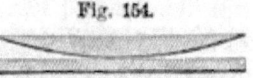

Fig. 154.

if different circles be described around this point,
at all parts of each circle the two surfaces will be
the same distance apart, and the larger the circle
the greater will be the distance. Now let a beam
of red light fall upon the flat surface. A black
spot is seen at the centre; around this a circle
of red light, then a dark ring, then another circle
of red light, and so alternating to the circumference. By careful measurement it is found that
the distances between the surfaces of the glass

where the circles of red light appear, are as the numbers 1, 2, 3, &c. This, taken in connection with what we know already of the theory of wave-motion, suggests at once the cause. There are two sets of waves, one reflected from the upper surface of the plane glass, and the other from the lower surface of the convex glass. Where their distance apart is less than a wave-length, they interfere and produce darkness. Where it is 1, 2, 3, or some whole-number of wave-lengths, they coalesce and produce a wave of greater intensity. To determine the length of a wave of red light, we have only to measure the distance between the two glasses at the first ring.

When beams of light of the various colors are used, circles of a corresponding color are obtained, and, singularly enough, the circles are of different diameters; red light gives the largest, and violet the smallest. We hence conclude that red waves are the longest, and violet the shortest.

Length of the Waves.—The minuteness of these waves passes comprehension. 40,000 red waves and 60,000 violet ones are comprised within a single inch. Knowing that light moves at the rate of 183,000 miles per second, we can easily calculate the number of these tiny waves which reach our eye in that time. When we look at a red object, 414 million million of ether waves break on the retina every moment, and with a violet color the number reaches 666 million million!

COLOR is exactly analogous to *pitch*. Violet corresponds to the high and red to the low sounds in music. *Intensity* of color, like that of sound, depends on the amplitude of the vibrations. When a body absorbs all the colors of the spectrum except blue, but reflects that to the eye, we call it a blue body; when it absorbs all but green, we call it a green body. Red glass has the power of absorbing all except the red rays, which it transmits. When a substance *reflects* all the colors to the eye, it seems to us white. If it *absorbs* all the colors, it is black. A *tint* is produced by a mingling of waves of different colors. We thus see that color is not an inherent property of the objects around us. In the darkness all bodies are devoid of color.

The play of colors seen in mother-of-pearl is due to the interference of light in the fine grooves caused by the edges of the thin overlapping plates of which it is composed. The same effect may be seen in a putty mould of the pearl. In a similar manner the plumage of certain birds reflects changeable hues. A metallic surface ruled with fine parallel lines not more than $\frac{1}{2000}$ of an inch apart, gleams with brilliant colors. Thin cracks in plates of glass or quartz, mica, when two layers are slightly separated—even the scum floating in stagnant water, break up the white light of the sunbeam and reflect the varying tints of the rainbow. The rich coloring of a soap-bubble is given it by the film of water which runs from the top down the sides, and thus produces interference of light.

DIFFRACTION OF LIGHT is caused by a beam of light passing along the edge of some opaque body. As the waves of ether strike against it, they put in motion another set of waves on the opposite side which interfere with the first system. If we hold a fine needle close to one eye and look toward the window, we shall see several needles. Place the blades of two knives closely together and hold them up to the sky; a most beautiful set of waving lines of interference will shade the open space. Most delicate colors are seen by looking at the sky through the meshes of a veil, or at a lamp-light through a bird-feather or a fine slit in a card.

POLARIZED LIGHT.—*Double Refraction.*—If we could 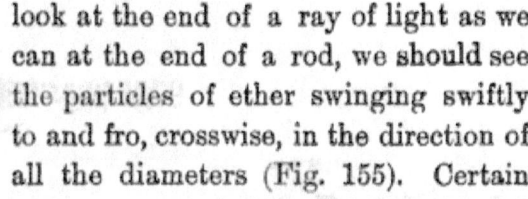 look at the end of a ray of light as we can at the end of a rod, we should see the particles of ether swinging swiftly to and fro, crosswise, in the direction of all the diameters (Fig. 155). Certain crystals have the power of sifting and arranging these vibrations into two sets at right angles to each

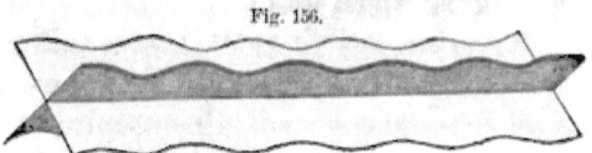

Fig. 156.

other, thus making a ray of the form seen in Fig. 156. As one set is more refracted than the other in passing through the crystal, the ray is divided into two rays—the *ordinary* and *extraordinary*. Rays which have been thus sifted constitute polarized light. Iceland spar (Fig. 157) possesses this property of double refraction in a remarkable degree. An object viewed through it in any direction not parallel to *a b* (the optic axis) appears double. If the crystal is

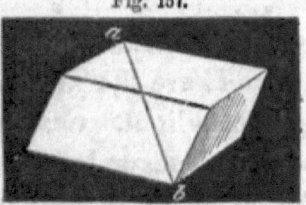

Fig. 157.

placed on a dot and slowly turned around, two dots will be seen, the second revolving about the first.

Light may be polarized by *reflection* from glass at a fixed angle, and also by passing through a thin slice of *tourmaline*, — a transparent crystal which absorbs the ordinary and transmits the extraordinary rays.

Fig. 158.

Object seen through Iceland Spar.

Objects examined by means of polarized light present many curious changes. A crystal of quartz or mica, which appears to the eye like common glass, reveals a series of beautifully colored rays, due to the interference of the ordinary and extraordinary rays of light. If we look at a lamp-light through a piece of common isinglass, we shall see a beautiful series of polarized rays having a star-like form. The angle of the rays varies with different kinds of mica. If polarized light be passed through common glass no change is seen, but on slight pressure a system of variegated colors appears. This method of examination presents a most delicate means of determining the molecular structure of a body. Some substances have the power of twisting the plane of the polarized light. Grape-sugar turns it to the right, and fruit-sugar to the left. The French Government use a polarizing instrument, in which this principle is applied to test the quality of the sugar imported into France.

THE RAINBOW is formed by the *refraction* and *reflection* of the sunbeam in drops of falling water. The white light of the sun is thus decomposed into its simple colors. The inner arch is termed the primary bow; the outer or fainter arch, the secondary. Each of these contains all the colors of the spectrum, but in reverse order. The rainbow is always seen in the quarter of the heavens opposite to the sun.

Primary Bow.—A ray of light, S″, enters, and is bent downward at the top of a falling drop, passes to the opposite side, is there reflected, then passing out of the lower side, is bent upward. By the refraction the ray of white light is decomposed, so that when it emerges it is spread out fan-like, as in the solar spectrum. Suppose that the eye of a spectator is in a proper position to receive the red ray, he cannot receive any other color from the same drop, because the red is bent upward the least, and all the others will pass directly over his head. He sees the violet in a drop below. Intermediate drops furnish the other colors of the spectrum.

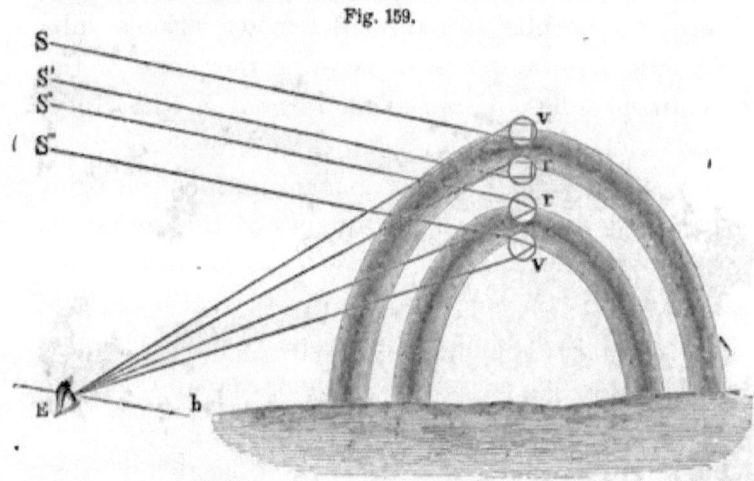

Fig. 159.

Secondary Bow.—A ray of light, S, strikes the bottom of a drop, v, is refracted upward, passes to the opposite side, where it is twice reflected, and thence passes out at the upper side of the drop. The violet ray being most refracted, is bent down to the eye of the spectator. Another drop, r, refracting another ray of light, is in the right position to send the red ray to the eye.

Why the Bow is circular.—In the primary bow it is found that when the red ray leaves the drop, it forms an angle with the sun's ray, S r, of about 42°, the violet 40°. These angles are constant. Let $a\,b$ be a straight line drawn from the sun through the observer's eye. If produced, it would pass through the centre of the circle of which the rainbow is an arc. This line is termed the *visual axis*. It is parallel to the rays of the sun; and when it is also parallel to the horizon, the rainbow is an exact semicircle. Suppose the line E v in the primary bow to be revolved around E b, keeping the angle b E v unchanged; the point v would describe a circle on the sky, and every drop over which it would pass would be at the proper angle to send a violet ray to the eye at E. Imagine the same with the drop r. We can thus see (1) that the bow must be circular; (2) that when the sun is high in the heavens, the whole bow sinks below the horizon; (3) that the lower the sun the larger is the visible circumference; and (4) that on lofty mountains a perfect circle may sometimes be seen.

Halos, coronas, sundogs, circles about the moon, the gorgeous tinting at sunrise and sunset, are all produced by the refraction and reflection of the sun's rays when passing through the clouds in the upper regions of the atmosphere. The phenomenon familiarly known as the "sun's drawing water," consists merely of the long shadows of broken clouds.*

SPHERICAL ABERRATION.—Rays which pass through a lens near the edge are brought to a focus sooner than those nearer the centre. Therefore, when an

---

* Spectrum analysis, twilight, and other kindred topics, are best understood in their relations to Astronomy. See "Fourteen Weeks in Astronomy."

image is clear around the edge, it will be indistinct at the centre, and *vice versa*. This wandering of the rays from the focus is termed spherical aberration.

Chromatic aberration is caused by the different refrangibility of the several colors which compose white light. The violet, being bent most, tends to come to a focus sooner than the red, which is bent least. This causes the play of colors seen around the image produced by an ordinary glass. It is remedied by using a second lens of different dispersive power, which counteracts the effect of the first. (Fig. 160.) Such a lens is said to be *achromatic* (colorless).

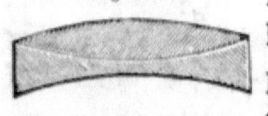
Fig. 160.

Optical Instruments.

MICROSCOPES (*to see small things*) are of two kinds, *simple* and *compound*. The former consists of a double convex lens; the latter contains at least two lenses.

At M is a mirror which reflects the rays of light through the object *a*. The object-lens *o* forms, in the tube above, a magnified inverted image of the object. The eye-lens O magnifies this image. If a microscope increases the diameter of an object 100 times, it is said to have a power of 100 *diameters*. In that case the surface is magnified $100^2 = 10,000$ times. To prevent spherical aberration the object-lens is made very small.

Fig. 161.

A Compound Microscope.

TELESCOPES (*to see afar off*) are of two kinds, *reflecting* and *refracting*. The former contains a large metallic mirror (speculum) which reflects the rays of light to a focus. The observer stands at the side and examines the image with an eye-piece.

OPTICS

Fig. 162.

A Reflecting Telescope.

The largest reflecting telescope ever made is that of Lord Rosse. Its speculum has a diameter of 6 feet, and a focal distance of 53 feet. (See frontispiece of Astronomy.)

The refracting telescope, like the microscope, contains an object-lens *o* which forms an image *a b*. This is viewed by means of the eye-piece O, which produces a magnified and inverted image *cd*. The objects seen in the heavens are so far distant that the rays of light are nearly parallel, and hence there is little spherical aberration. The object-lens may there-

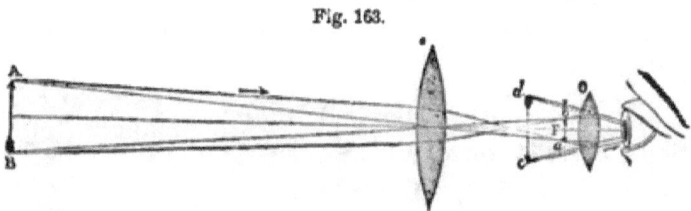

Fig. 163.

fore be made of any size without rendering the image indistinct. The larger this lens, the more light is collected with which to view the image. The magnifying power is principally due to the eye-piece. The great telescope in the observatory at Chicago is the best in the world. The diameter of its object-glass is $18\frac{1}{2}$ inches—equivalent to enlarging the pupil of the astronomer's eye to that size. It was made by Alvan Clark & Sons, of Cambridge, Massachusetts.

The inversion of the object is of no practical importance for astronomical purposes. For terrestrial observations additional lenses are used to invert the image.

The opera-glass contains an object-glass O and an eye-piece *o*. The latter is a double concave lens; this increases the visual angle by diverging the rays

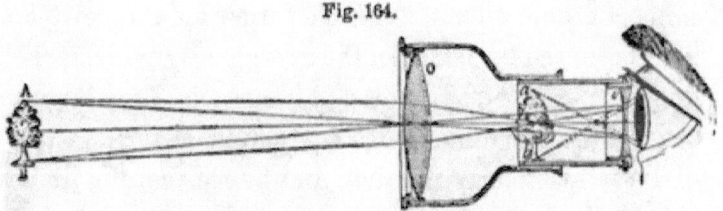

Fig. 164.

of light, which would otherwise come to a focus beyond the eye-piece. An erect and magnified image is seen at $a\,b$.

The stereoscope contains portions of two convex lenses, as shown in Fig. 165. Two photographs A and B are taken by two cameras which are slightly inclined to each other. This produces two pictures precisely like the two views we always obtain of an object by the use of both eyes. The blending of the two at C causes the appearance of solidity.

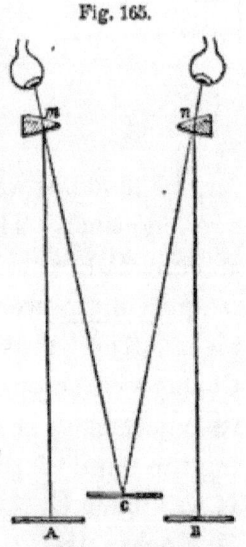

Fig. 165.

The magic lantern, or stereopticon, contains a reflector M, which condenses the rays of a powerful oil or calcium light upon a lens L. They are here converged upon the object $a\,b$. Thence a double lens m throws a highly magnified image on the screen A B. *Dissolving views* are produced by the use of two lanterns which contain the separate scenes which are to melt into each other.

Fig. 166.

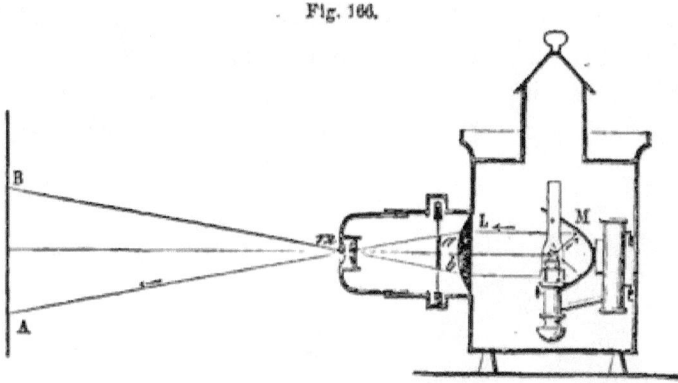

The Camera used by photographers contains a double-convex lens at A, which throws an inverted

Fig. 167.

image of the object upon the ground-glass screen E B.

THE EYE is the most perfect optical instrument. It is rarely, if ever, troubled by spherical or chromatic aberration, and is self-focusing. It closely resembles a camera. The outer membrane of the eye is termed the *Sclerotic* coat, S. It is tough.

white, opaque, and firm. A little portion in front, termed the *Cornea*, c, is transparent; this is convex, and is set into the sclerotic coat somewhat like a watch-crystal. The middle or *Choroid* coat, C, is soft

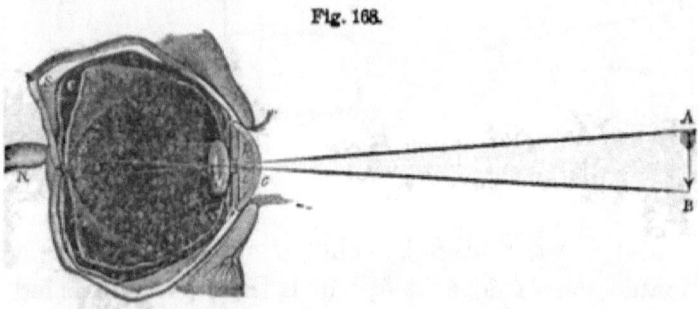

Fig. 168.

and delicate, like velvet. It lines the inner part of the eye. It is covered with a black pigment, which absorbs the superfluous light. Upon it the optic nerve, which enters at the rear, expands in a network of delicate fibres termed the *Retina*. This is the seat of vision. Back of the cornea is a transparent limpid fluid, the *aqueous humor*. The *anterior* chamber, filled with this liquid, is closed at the back by a colored curtain, h i, the *Iris*. The *Pupil* is a round hole in the Iris. The *Crystalline lens*, o, separates the two chambers of the eye. It is a double-convex lens, tough as gristle and transparent as glass. It is composed of concentric layers, like an onion. The *posterior* chamber is filled with the *vitreous humor*, which is a transparent, jelly-like liquid, resembling the white of an egg.

Let A B represent an object in front of the eye.

Rays of light are first refracted by the aqueous humor, then falling upon the crystalline lens they are further refracted, and lastly are refracted by the vitreous humor and form an image $a\,b$ on the retina. This is smaller than the object, and inverted.

The more distant the object, the smaller the picture. To render vision distinct, the rays must be accurately focused on the retina. If we gaze steadily at an object near by, and then suddenly observe some distant one, we find our vision blurred. In a few moments it becomes clear again. This shows that the eye has the power of adapting itself to the varying distances of objects, which is done by a variation in the convexity of the crystalline lens.

When a body is held very near the eye, the lens has not sufficient power to converge the rays upon the retina in a distinct image. When the distance at which a clear vision takes place is less than four or five inches, the person is said to be *near-sighted*, and when greater than ten or twelve, to be *far-sighted*. This difference lies in the shape of the eyeball. In far-sightedness the ball is too flat, and the retina is too near the lens; in near-sightedness the ball is elongated, so that the retina is too far back from the lens. The former can be remedied by convex glasses, which bring the rays to a focus sooner, and the latter by concave glasses, which throw the focus further back. In old age the eye loses the power of adjusting the crystalline lens; elderly people, therefore, hold a book at some distance from the eye. They are aided by using convex glasses.

The retina retains any impression made upon it for about one-eighth of a second. This explains why a wheel, when rapidly revolved, appears solid, or a lighted brand like a ring of fire. On the other hand, it requires a moment for an impression to be made. Thus a wheel may be whirled so swiftly that its spokes become invisible.

Some eyes are entirely uninfluenced by certain colors, as some ears are deaf to certain sounds. Color-blindness is most commonly noticed in reference to red, green, and blue. Doubtless railway accidents have often occurred through this natural inability to distinguish signals. Dr. Mitchell mentions the case of a naval officer who chose for his uniform a blue coat and red waistcoat, fully believing them to be of the same color. He also tells of a tailor who mended a black silk waistcoat with a piece of crimson, and of another who put a red collar on a blue coat. Dalton could only see two colors, blue and yellow, in the solar spectrum, and having once dropped a piece of red sealing-wax in the grass, he could not find it by the difference in color. The range of the eye is much less than that of the ear. The latter is about eleven octaves, while the former never exceeds a single octave.

The diameter of the eye is less than an inch; yet, as we look over an extended landscape, every feature, with all its variety of shade and color, is repeated in miniature on the retina. Millions upon millions of ether waves, converging from every direction, break on that tiny beach, while we, oblivious to all the marvellous nature of the act, think only of the beauty of the revelation.

Practical Questions.—**1.** Why is the secondary bow fainter than the primary? Why are the colors reversed? **2.** Why can we not see around the corner of a house, or through a bent tube? **3.** What color would a painter use if he wished to represent an opening into a dark cellar? **4.** Is white a color? Is black? **5.** By holding an object nearer a light, will it increase or diminish the size of the shadow? **6.** What must be the size of a glass in order to reflect a full-length image of a person? *Ans.* Half the person's height. **7.** Where may we see a rainbow in the morning? **8.** Can two spectators see the same bow? **9.** Why, when the drops of water are falling through the air, does the rainbow appear stationary? **10.** Why can a cat see in the night? **11.** Why cannot an owl see in daylight? **12.** Why are we blinded when we pass quickly from a dark into a brilliantly lighted room? **13.** If the light of the sun upon a distant planet is only $1/100$ of that which we receive, how does its distance from the sun compare with ours? **14.** If when I sit six feet from a candle I receive a certain amount of light, how much shall I diminish it if I move back six feet further? **15.** Why do drops of rain, in falling, appear like liquid threads? **16.** Why does a towel turn darker when wet? **17.** Does color exist in the object, or in the mind of the observer? **18.** Why is lather opaque, while air and a solution of soap are each transparent? **19.** Why does it whiten molasses candy to "pull it?" **20.** Why does plastering become lighter in color as it dries? **21.** Why does the photographer use a kerosene oil lamp in the "dark room?" **22.** Is the common division of colors into "cold" and "warm" verified in philosophy? **23.** Why is the image on the camera, Fig. 167, inverted? **24.** Why is the second image seen in a mirror, Fig. 134, brighter than the first? **25.** Why does a blow on the head make one "see stars?" *Ans.* The blow excites the optic nerve, and so produces the sensation of light. **26.** What is the principle of the kaleidoscope? *Ans.* It contains three mirrors set at an angle of 60°. Small bits of colored glass at one end reflect to the eye, at the other, multiple images which change in varying patterns as the tube is revolved. **27.** Which can be heard at the greater distance, noise or music?

HEAT.

DEFINITIONS.—*Luminous* heat is that which radiates from a luminous body. Ex.: An iron heated to whiteness. *Obscure* heat is that from a non-luminous source. A *diathermanous* (*dia*, through, and *thermos*, warm) body is one which allows the heat to pass through it readily. Rock-salt is the most perfect diathermanous solid known. It is to heat what glass is to light. *Cold* is a merely relative term, indicating the absence of heat in a greater or less degree. *Gases* and *Vapors* differ but slightly. The former retain their form at all ordinary temperatures; Ex.: Air. The latter are readily condensed; Ex.: Steam.

THE INTIMATE RELATION BETWEEN LIGHT AND HEAT.—Thrust a cold iron into the fire. It is at first dark, but soon becomes luminous, like the glowing coals. Raise the temperature of a platinum wire. We soon feel the radiation of obscure heat-rays. As the metal begins to glow, our eye detects a red color, then orange combined with it, then green, and so on through the scale of the spectrum. At last all the colors are emitted, and the metal is dazzling white. All bodies become luminous at a fixed temperature. Heat may be reflected, refracted, and even polarized. It radiates in straight lines equally in every direction, and decreases in intensity as the square of the distance. It moves with the same velocity as light. Heat and light come to us combined in the sunbeam. It is therefore believed that they are the same—that light is only luminous heat, and that the three classes of waves in the solar spectrum differ, as one color differs from another, in the rapidity of the vibrations. The longer and slower waves of ether fall upon the nerves of touch, and produce the sensation of heat.* The more rapid ones are peculiarly adapted to affect the optic nerve and produce the sensation of light. The shortest and quickest cause chemical changes.

THEORY OF HEAT.—Heat is only a mode of motion. Rest is unknown in nature. Even the molecules of

---

* It is now believed that the particles of the nerves vibrate, and thus communicate to the brain the impressions made by external objects. Each of the five classes of nerves seems to be adapted to transmit vibrations of its own kind, while it is insensible to all others. Thus, if the rate of oscillation be less than that of red, or more than that of violet, the optic nerve is uninfluenced by the waves. We cannot see with our fingers, nor taste with our ears. Nerves transmit motion at the rate of about 93 feet per second. If, then, a man, six feet high, were to step on a nail, it would require nearly an eighth of a second for the information to be carried by the sensor nerve to the brain, and for the order to lift the foot to be returned by the motor nerve to the suffering member.

a solid are in constant vibration. As the worlds in space are revolving about each other in inconceivably *vast* orbits, so each body forms a miniature system, its molecules revolving in inconceivably *minute* orbits. When we *increase* the rapidity of this motion, we heat the body; when we *decrease* it, we cool the body. The vacant spaces between the molecules are filled with ether. As the air moving among the limbs of a tree sets its boughs in motion, and in turn may be kept in motion by the waving branches, so this ether may put the molecules in vibration, or be thrown into motion by them. Ex.: Insert one end of a poker into the fire. The particles in contact with the heat are made to vibrate intensely; these swinging atoms strike their neighbors, and so on, atom by atom, until the oscillation reaches the other end. If, now, we handle the poker, the motion is imparted to the delicate nerves of touch; they carry it to the brain, and pain is felt. In popular language, "the iron is hot," and we are burned. If, without touching it, we hold our hand near the poker, the ether waves set in motion by the whirling atoms of iron strike against the hand, and produce a less intense sensation of heat. In the former case, the fierce motion is imparted directly; in the latter, the ether acts as a carrier to bring it to us.

QUALITY OF HEAT.—As some sounds are shrill and piercing, others deep and heavy, so some kinds of heat are keen and penetrating, others mild and diffusive. This difference depends on the length of the waves and their combined rates of vibration. Thus the chirp of the cricket compares with the heat of a glowing furnace, and the soft tones of an organ with the genial radiation of a steam-pipe. Pitch in music, variety in color, and degrees in heat, are therefore intimately related.

THE SOURCES OF HEAT are the sun, stars, mechanical and chemical forces.

(1.) The molecules of the sun and stars are in rapid vibration. These set in motion waves of ether, which dart with the velocity of light across the intervening space, and surging against the earth, give up their motion to it. (2.) Friction and percussion produce heat, because additional motion is thereby imparted to the vibrating particles. Savages obtain fire by rubbing together two pieces of wood. A horse hits his steel shoes against a stone and "strikes fire;" little particles of the metal torn off are heated by the shock, so that they burn as sparks. The bearings of machinery become hot, unless the friction is diminished by grease. A train of cars is stopped by the pressure of the brakes. If we watch them in a dark night, we shall see the sparks flying from the wheels, the motion of the train being converted into heat. A blacksmith pounds a piece of iron until it glows. The force of his strokes sets the particles of the metal vibrating rapidly enough to send ether waves of such swiftness as to affect the eye of the observer.* A piece of wood may be heated by simply squeezing it in a hydraulic press. At the exposition in Paris, chocolate was kept hot by means of two copper plates which were rubbed together by machinery. As a cannon-shot strikes an iron target, a sheet of flame pours from it. Our earth moves with a velocity of over 68,000 miles per hour. Were it instantly stopped, enough heat would be produced to change the entire globe to vapor. (3.) Chemical action is most commonly seen in fire. The oxygen of the air has an affinity for the carbon and hydrogen of the fuel. They rush together. As they strike, their motion is stopped. The shock sets the molecules in vibration. They impart their motion to the ether, and thus start waves of heat.

THE MECHANICAL EQUIVALENT OF HEAT.—In these various changes of mechanical-motion into heat-motion no force has been lost. The blacksmith's hammer falling on the anvil gives rise to a definite amount of heat. If the heat could be gathered up, it would be sufficient to lift the hammer to the point from which it fell. No force can be annihilated. If destroyed in one form, it reappears in another without loss. *A pound-weight falling from a height of 772 feet, would generate enough heat to raise the*

---

* Text-books frequently assert that "iron, once treated in this way, cannot again be made red-hot by hammering, unless subsequently reheated in the forge." Even Miller gives currency to this statement. Any blacksmith will convince you of its utter falsity by actual experiment.

temperature of one pound of water one degree; conversely, the amount of heat necessary to elevate a pound of water one degree, would raise a pound-weight to the height of 772 feet. This is called "*Joule's law*," or the "mechanical equivalent of heat."

Change of State by Heat.

Latent, Sensible, and Specific Heat.—When a body is heated, the heat-force is divided into two parts: one portion elevates the temperature, and the other increases the size. The former can be detected by the touch, and is called *sensible* heat. The latter tends to counteract the force of Cohesion, and is neutralized so that it cannot be detected by the touch; it is therefore termed *latent* heat. Substances vary in their application of the heat-force. Some devote more to temperature, others to **expansion**. Therefore, if the same amount of heat be applied to the same bulk of different substances, they will not indicate the same temperature; and on the other hand, when various bodies indicate the same temperature by a thermometer, they may possess vastly different quantities of heat. Steam contains the greatest amount of latent heat **of any known** substance, except hydrogen, **yet it indicates** no higher temperature than boiling water. The relation between sensible and latent heat is termed *specific* heat. It is the quantity of heat that is required to raise a given weight of any substance 1° in temperature, compared with the quantity required to elevate the same weight of water 1°. Thus, a quantity of heat which elevates the temperature of a pound of water 1° would raise that of a pound of mercury 30°. Hence, taking water as the standard, the specific heat of mercury is only .033.

Latent heat is not lost.—In the various changes of state of which we shall now speak, wherein bodies pass from solid to liquid and from liquid to the gaseous form, sensible heat becomes latent. Thus, one who has melted snow, or "boiled away" water, knows how slow is the process, and how much heat is consumed. When the vapor or liquid passes back into its original state, the latent heat is restored again as sensible heat. The following curious paradox illustrates this thought: *Freezing is a warming process, and thawing a cooling process.*

Freezing-mixtures depend on the principle of latent heat. Their most common use is in freezing ice-cream. Salt and powdered ice are employed. Salt has a great attraction for water. It therefore dissolves the ice to get it, and then itself dissolves in the water thus obtained. In this process two solids pass into the liquid form. The heat necessary for this change of state is absorbed mainly from the cream.

I. EXPANSION.—By the addition of heat the molecules are urged into swifter motion, and therefore pushed further apart, increasing the size of the body. Hence the law, "*Heat expands and cold contracts.*"

(1.) *Solids* expand uniformly; *i. e.*, a definite rise

in temperature produces a fixed increase of size. Different substances, however, expand unequally. Zinc dilates more than iron, and iron more than glass. The force of the expansion is irresistible. It is said that iron, heated from zero up to the boiling point of water, exerts a pressure equal to 16,000 times that of gravity. When it cools, it contracts with the same force. Practical applications of this principle abound in the arts. A carriage-tire is put on when hot, that, when cooled, it may bind the wheel together. Rivets used in fastening the plates of steam-boilers are inserted red-hot. "The ponderous iron tubes of the Britannia bridge writhe and twist, like a huge serpent, under the varying influence of the solar heat. A span of the tube is depressed but a quarter of an inch by the heaviest train of cars, while the sun lifts it $2\frac{1}{2}$ inches." The Bunkerhill monument nods as it follows the sun in its daily course. Lead and zinc, on cooling, do not contract to their original dimensions, but their particles slide over each other in expanding; in this manner the linings of sinks become puckered. On the other hand, this force of expansion must be guarded against. In laying long water-pipes, some of the tubes are made to slip into each other with telescopic slides. Tumblers of thick glass often break on the sudden application of heat, because the surface dilates before the motion has time to reach the interior. Draughts of cold air frequently crack heated lamp-chimneys, for a similar reason. (2).

Liquids are much more sensitive to heat than solids, but do not expand as equally. (3). *Gases* expand uniformly $\frac{1}{490}$ of their bulk. 490 cubic inches of any gas, at 32° F., if heated 1°, become 491 cubic inches.

The Mercurial Thermometer is an instrument for measuring the temperature by the expansion of mercury. As this metal freezes at —39° F., colored alcohol is used for low temperatures. The method of filling a thermometer may be illustrated in the following manner. Take the glass tube shown in Fig. 67, and hold the bulb in the flame of an alcohol lamp until the air is nearly expelled. Then plunge the stem in some colored water. As soon as the bulb cools, the water will rise and partly fill it. Heat the bulb again in the flame until the steam pours out of the stem. On inserting it a second time, the water will entirely fill the bulb. In the manufacture of thermometers, it is customary to have a cup blown at the upper end of the stem. This is filled with mercury, and the air, when expanded, bubbles out through it, while the metal trickles down as the bulb cools. The mercury is now heated to as high a temperature as the thermometer is intended to measure, when the tube is melted off and sealed at the extremity of the column of mercury. The metal contracts on cooling, and leaves a vacuum above. Each thermometer is graduated separately. It is put in melting ice, and the point to which the mercury sinks is marked 32°—*Freezing-point*. It is then placed in a steam-bath, and the point to which the mercury rises (when the barometric column stands at 30 in.) is marked 212°—*Boiling-point.* This constitutes what is called Fahrenheit's scale (F.) It is said that the inventor placed the zero-point 32° below the temperature of freezing water, because he thought that to be absolute cold. In the Centigrade scale (C.), the freezing-point is marked 0, and the boiling-point 100. 1° C. = 1.8°F. In Reaumur's scale (R.), the boiling-point is fixed at 80°.

Fig. 169.

II. LIQUEFACTION.—When heat is added to a solid body, the temperature rises until the freezing-point (melting-point) is reached, when it becomes stationary. The force is now all applied to neutralizing the Cohesive attraction. The expansion continues. The molecules are pushed further and further apart, until, escaping the grasp of Cohesion, they move freely on each other. This constitutes liquefaction, as seen in the melting of ice, iron, &c. In this process large quantities of heat become latent in the body. If ice at 32° be melted, 142° of heat will disappear, and the water will be at only 32°. Hence, to convert 1 lb. of ice at 32° into water at 32°, enough heat must be used to raise 142 lbs. of water from 32° to 33°.

Liquefaction of gases.—When a gas is cooled the repellant force is weakened, and the molecules once more approach each other. By the combined action of cold and pressure the particles of almost every known gas have been brought near enough for the attraction of Cohesion to grasp them. When the pressure is removed, the gaseous form is quickly resumed.

III. VAPORIZATION.—If heat be applied to a liquid, the temperature rises until the *boiling-point* is reached, when it stops. The expansion, however, continues until the motion is so violent as to overcome the Cohesive force and to throw off particles of the liquid.

Fig. 170.

The vapor thus formed does not contain any solid which may be dissolved in the liquid. This principle is applied to *distillation*. Ex.: Pure or distilled water is obtained by heating it in a boiler A, whence the steam passes through the pipe C and the *worm* within the condenser S, where it is condensed and drops out into the vessel D. The pipe is coiled in a spiral form within the condenser, and is hence termed the worm. The condenser is kept full of cold water by means of the tub at the left. By careful regulation of the heat, one liquid may be separated from another by distillation. (See Chemistry, p. 196.)

The boiling-point.—When we heat water, the bubbles which pass off first contain merely the air dissolved in the liquid; next bubbles of steam form on the bottom and sides of the vessel, and, rising a little distance, are crushed in by the cold water and condensed. In breaking they produce that peculiar sound known as "simmering." They ascend higher and higher as the **temperature** of the water rises, until at last they break at the surface, and the steam passes off into the air. The violent agitation of the water produced by the passage of these steam bubbles is termed boiling. The boiling-point is not the same in different liquids. This produces the variety we see in the forms of matter. Some vaporize at ordinary temperatures; others only melt at the very highest; while the gases of the air are but the steam of substances which **vaporize** at enormously low temperatures. The boiling-point of water depends on three circumstances.

(1.) *The purity of the water.*—Any substance which increases the cohesive power of the water elevates the boiling-point. For this reason salt water boils at a higher temperature than pure water. The air dissolved in water tends by its elastic force to separate the molecules. If this be removed, the boiling-point is elevated as high even as 275°, when the water is converted into steam with explosive violence.

(2.) *The nature of the vessel.*—Water will boil at a lower temperature in an iron than in a glass vessel. If the surface of the glass be made chemically clean, the boiling-point is elevated still higher. This seems to depend on the strength of the adhesion between the water and the vessel in which it is contained.

(3.) *The pressure.*—Any pressure upon the surface tends to keep the molecules together, and so raises the boiling-point. Water, therefore, boils at a lower temperature on a mountain than in a valley. The temperature of boiling water at Quito is 90°, and on Mont Blanc 84° C. The variation is so uniform, that the height of any place can be ascertained with tolerable accuracy by this means. A difference of 1° F. is produced by an ascent of 596 feet.

The influence of pressure is very finely illustrated by the following experiment. Boil a glass flask half full of water for some time. Cork it quickly and then invert it. The pressure of the accumulated steam will soon stop all ebullition. A few drops of cold water will condense the steam, and boiling will commence again. This will soon be checked, but can be restored as before. The process may be repeated until the water cools to little more than blood-heat.

Fig. 171.

If the cork be airtight, when the water is quite cold it will strike with a sharp metallic sound as it falls from one end of the flask to the other. The cushion of air which commonly breaks the fall of water is here removed. The *water-hammer* illustrates this point yet more fully. It consists of a glass tube half full of water, from which the air has been expelled by heat, the tube being sealed while the water was yet boiling. The vacuum is very perfect; steam may be produced in it by the

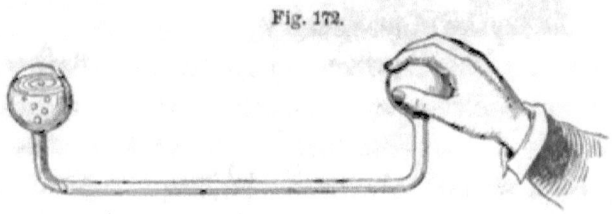

Fig. 172.

heat of the hand, and the water falls to and fro with the apparent force of lead. The pulse-glass shown in the figure is a somewhat similar instrument.

The temperature cannot be raised above the boiling-point, unless the steam is confined, however much heat may be applied. The extra force is entirely occupied in expanding the water into steam. This occupies 1,700 times the space, and is of the same temperature as the water from which it is made. Over 900° of heat become latent in this process, but are made sensible again when the steam is condensed. The common method of heating by steam depends upon this fact. Steam is invisible. This we can verify for ourselves by examining it just as it issues from the spout of the tea-kettle. It soon condenses, however, into minute globules, which, floating in the true steam, render the vapor apparently visible.

IV. EVAPORATION should be distinguished from vaporization. It is a slow formation of vapor, which takes place at all ordinary temperatures. **Ex.**: Water evaporates slowly, even at the freezing-point. Clothes dry in the open air in the coldest weather. The wind quickens the process, because it drives away the moist air near the clothes and supplies its place with dry air. Evaporation is also hastened by an increase of surface and a gentle heat. This principle is made useful in the arts for separating a solid from the liquid which holds it in solution.

Vacuum pans are largely employed in condensing milk, in the manufacture of sugar, etc. They are so arranged that the air above the liquid in the vessel may be exhausted, and then the evaporation takes

Fig. 171.

If the cork be airtight, when the water is quite cold it will strike with a sharp metallic sound as it falls from one end of the flask to the other. The cushion of air which commonly breaks the fall of water is here removed. The *water-hammer* illustrates this point yet more fully. It consists of a glass tube half full of water, from which the air has been expelled by heat, the tube being sealed while the water was yet boiling. The vacuum is very perfect; steam may be produced in it by the

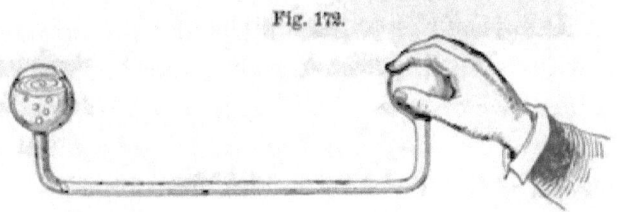

Fig. 172.

heat of the hand, and the water falls to and fro with the apparent force of lead. The pulse-glass shown in the figure is a somewhat similar instrument.

The temperature cannot be raised above the boiling-point, unless the steam is confined, however much heat may be applied. The extra force is entirely occupied in expanding the water into steam. This occupies 1,700 times the space, and is of the same temperature as the water from which it is made. Over 900° of heat become latent in this process, but are made sensible again when the steam is condensed. The common method of heating by steam depends upon this fact. Steam is invisible. This we can verify for ourselves by examining it just as it issues from the spout of the tea-kettle. It soon condenses, however, into minute globules, which, floating in the true steam, render the vapor apparently visible.

IV. EVAPORATION should be distinguished from vaporization. It is a slow formation of vapor, which takes place at all ordinary temperatures. Ex.: Water evaporates slowly, even at the freezing-point. Clothes dry in the open air in the coldest weather. The wind quickens the process, because it drives away the moist air near the clothes and supplies its place with dry air. Evaporation is also hastened by an increase of surface and a gentle heat. This principle is made useful in the arts for separating a solid from the liquid which holds it in solution.

Vacuum pans are largely employed in condensing milk, in the manufacture of sugar, etc. They are so arranged that the air above the liquid in the vessel may be exhausted, and then the evaporation takes place very rapidly, and at so low a temperature that all danger of burning is avoided.

Fig. 173.

Evaporation cools a liquid very rapidly, since sensible heat becomes latent in the vapor. Pious Mahometans were formerly accustomed to place, in niches along the public streets, porous earthenware bottles filled with water, to refresh the thirsty travellers. Water may even be frozen in a vacuum, if the vapor be removed as fast as formed. Ice is manufactured in the tropics by machines constructed on this principle. The greatest artificial cold ever known, $-220°$ F., was produced by evaporating in a vacuum a mixture of liquid nitrous oxide gas and bisulphide of carbon.

SPHEROIDAL STATE.—If a few drops of water be put in a red-hot metallic cup, they will gather into a globule, which will dart to and fro over the surface with little diminution. It seems to rest on a little cushion of steam, which supports it while the heated currents of air drive it hither and thither. If the cup be allowed to cool, after a little the water will lose its spheroidal form, and coming into direct contact with the metal, burst into steam with a slight explosion. This principle may perhaps account for some unexplained boiler accidents. By moistening our finger, we can touch a hot flat-iron with impunity. The water assumes the above state, and thus protects the flesh from injury. Furnace-men often dip their moistened hands into molten iron. Probably the accounts handed down to us of persons walking unharmed over red-hot ploughshares are to be explained in this manner.

COMMUNICATION OF HEAT.

Heat tends to diffuse itself equally among all surrounding bodies. There are three modes of distribution.

I. CONDUCTION *is the process of heating by the passage of heat from molecule to molecule.* Ex.: Hold one end of a poker in the fire, and the other end soon becomes hot enough to burn the hand. Substances vary in their power of conduction. The denser bodies, as the metals, possess this property in the highest degree. Of the ordinary metals, copper is the best conductor. Wood is a poor conductor, especially "across the grain." *Gases* are the poorest conductors; hence porous bodies, as wool, fur, snow, charcoal, etc., which contain within them large quantities of air, are excellent non-conductors. Refrigerators and ice-houses have double walls, filled between with charcoal, sawdust, or other non-conducting substances. Fire-safes contain plaster-of-paris. Air is so poor a conductor, that persons have gone into ovens, which were so hot as to cook meat and eggs which they carried in with them and laid on the metal shelves;

yet, so long as they did not themselves touch any good conductor, they experienced little inconvenience. *Liquids* are also poor conductors. Ex.: Hold the upper end of a test-tube of water in the flame of a lamp. The water nearest the blaze will boil, without the heat being felt by the hand.

All adjacent objects have the same temperature, yet flannel sheets feel warm, and linen cold. A marble slab seems colder than the woollen carpet below it. If we touch an object colder than we are, it abstracts heat from us, and we say "it feels cold;" if a warmer body, it imparts heat to us, and we say "it feels warm." These various effects depend entirely upon the relative conducting power of the different substances. Iron feels colder than feathers only because it robs us faster of our heat.

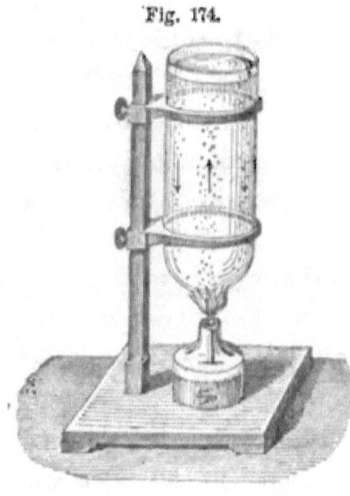

Fig. 174.

II. CONVECTION *is the process of heating by circulation.*—(1.) *Convection of liquids.* Place a little sawdust in a flask of water, and apply heat at the bottom. We shall soon find that an ascending and a descending current are established. The water near the lamp becoming heated, expands and rises. The cold water above sinks to take its place. (2.)

Convection of gases. By testing with a lighted candle, we shall find that at the bottom of a door opening into cold air, there is a current setting inward, and at the top, one setting outward. The cold air in a room flows to the stove along the floor, is heated, and then rises to the ceiling. All methods of heating by hot-air furnaces depend upon the principle that warm air rises.

III. RADIATION *is the process of heating by the transmission of rays in straight lines.* All heat from the sun comes to the earth in this manner. A hot stove radiates heat. Rays of heat do not elevate the temperature of the media through which they pass. When the motion of the ether-waves is stopped, the effect is felt. Space is not warmed by the sunbeam. Meat can be cooked by radiation, while the air around is at the freezing-point. Radiation varies in different bodies, and in the same body under different circumstances. A rough unpolished surface is a better radiator than a smooth bright one. Extent of surface increases radiation. Air, vapor, and glass allow *luminous* rays of heat to pass through them readily. Thus the heat of the sunbeam easily penetrates our atmosphere, windows, etc. But the earth, and various objects on its surface, absorb and radiate the heat back again as *obscure* heat in long, slow waves. These have no power to pass the watery vapor in the air or through glass. They are thus entangled, and kept for our use. If the aqueous vapor were removed from our air, the earth would become uninhabitable, through the rapid radiation. On the desert of Sahara, where "the soil is fire and the wind is flame," the dry air allows the heat to escape through it so readily that ice is sometimes formed at night. The dryness of the air at great elevations accounts, in part, for the coldness which is there felt so keenly.

Absorption and *reflection* are intimately connected with radiation. A good absorber is also a good radiator, but a good reflector can be neither. Snow is a good reflector but a poor absorber or radiator. Light colors absorb less and reflect more than dark colors. White is the best reflector, and black the best absorber and radiator.

THE STEAM-ENGINE.

When steam rises from water at a temperature of 212° it has an elastic force of 15 lbs. per square inch. If the steam is confined and the temperature raised, the elastic force is rapidly increased.

THE STEAM-ENGINE is a machine for using the elastic force of steam as a motive power. There are two classes of engines, the *high-pressure* and the *low-pressure*. In the former, the steam, after being employed to do its work, is forced out into the air; in the latter, it is condensed in a separate chamber by a spray of cold water. As the steam is condensed in the low-pressure engine, a vacuum is formed behind the piston; while the piston of the high-pressure engine acts against the pressure of the air.

The elastic force of the steam must be 15 lbs. per square inch greater in the latter case. In the figure we have represented the piston and connecting pipes of an engine. The steam from the boiler passes through the pipe into the steam-chest, as indicated by the arrow. The sliding-valve worked by the rod h lets the steam into the cylinder alternately above and below the piston, which is thus made to play up and down by the expansive force.

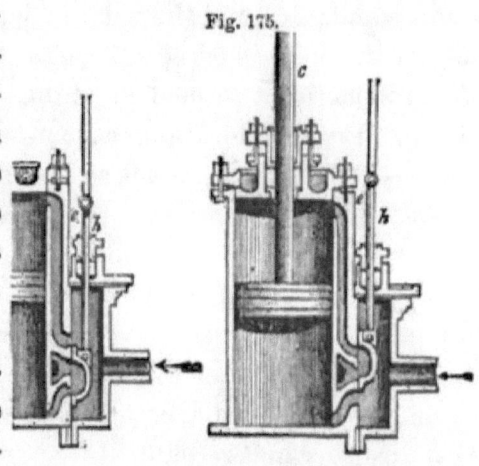

Fig. 175.

The Governor is an apparatus for regulating the supply of steam. A B is the axis around which the heavy balls E and D revolve. When the machine is going too fast the balls fly out by centrifugal force and shut off a portion of the steam; when too slowly, they fall back, and, opening the valve, let on the steam again.

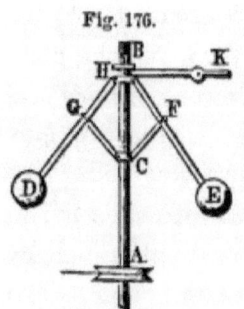

Fig. 176.

The high-pressure engine in the form commonly used is shown in the frontispiece. A represents the cylinder, B the steam-chest, C the throttle-valve in the pipe through which steam is admitted from the boiler, D the governor, E the band-wheel by which the governor is driven, F the pump, G the crank, I the connector which is attached to *a* the cross-head, H the eccentric rod (*h* in Fig. 175) which works the sliding-valve in the steam-chest, K the governor-valve, S the shaft by which the power is conveyed to the machinery. The cross-head, *a*, slides to and fro in a groove, and is fastened to the rod which works the piston in the cylinder A. The expansive force of the steam is thus communicated to *a*, thence to I, by which the crank is turned. The heavy fly-wheel, by its inertia, serves to render the movement of the machinery uniform.

METEOROLOGY.

The air always contains **moisture**. The amount it can receive depends on the temperature; warm air absorbing more, and cold air less. At 75° the vapor is sometimes so dense that in a cubic yard of atmosphere there is a cubic inch of water. At 50° half that quantity must be deposited. When the air at any temperature contains all the vapor it can hold, it is said to be *saturated; any fall of temperature will then cause a part of the vapor to be condensed.* Most of the phenomena of rain, hail, dew, etc., depend on this principle.

A change in density produces a change in temperature. Place a little tinder at the end of the piston of the fire-syringe shown in the figure. By forcing down the handle and compressing the air, sufficient heat is liberated to ignite the tinder. On the other hand, in experiments with the Air-pump we notice that as the air is rarefied, a mist gathers in the glass receiver. This shows that the atmosphere is cooled by its expansion, and so deposits its vapor. The warm air from the earth ascending into the upper regions, is rarefied and cooled in the same manner. Its vapor is condensed into clouds, and often falls as rain. Owing to this expansion of the air, there is a gradual diminution of the temperature as the altitude is increased, at the rate of about 1° for every 300 feet. Even in tropical climates the tops of high mountains are covered with perpetual snow. At the equator the snow-line is 15,000 feet above the level of the sea. Should, however, a blast of cold air descend from a lofty height, it would become so condensed in fall-

Fig. 177.

11*

ing, and its temperature thereby so elevated, that it would produce no injurious effect on vegetation.

DEW.—The grass at night, becoming cooled by radiation, condenses upon its surface the vapor of the air.* Dew will gather most freely upon those objects that are the best radiators, as they will the soonest become cool. Thus grass, leaves, etc., which need the most, get the most. It will not form on windy nights, because the air is constantly changing and does not become cool enough to deposit its moisture. In tropical regions the nocturnal radiation is often so great as to admit of the formation of ice. In Bengal this is accomplished by exposing water in shallow earthen dishes resting on rice-straw. The most dew collects on a clear, cloudless night. In many countries, by its abundance, it supplies the place of rain, as in Chili, Arabia, etc. When the temperature of plants falls below 32°, the vapor is frozen upon them directly, and is called *hoar-frost*.

FOGS are formed when the temperature of the air falls below the *dew-point* (*i. e.*, the temperature at which dew is deposited). They are found mainly

---

* Dew was anciently thought to possess many wonderful properties. Baths in this precious liquid were said to conduce greatly to beauty. It was collected for this purpose, and for the use of the alchemists in their weird experiments, by spreading fleeces of wool upon the ground. Laurens, a philosopher of the middle ages, claimed that dew is ethereal, so that if we should fill a lark's egg with it and lay it out in the sun, immediately on the rising of that luminary, the egg will fly off into the air! This experiment is best performed with a goose's egg.

on low grounds and in the vicinity of rivers, ponds, etc., where the abundance of moisture keeps the air constantly saturated.

Clouds differ from fogs only in their elevation in the atmosphere. They are formed when a "warm, humid wind penetrates a cold air, or a cold wind a warm, humid air." Mountains are "cloud-capped"

Fig. 178.

Different kinds of clouds—1 bird indicates the nimbus, 2 birds the stratus, 3 birds the cumulus, and 4 birds the cirrus cloud.

because the warm air rising from the valley is condensed upon their cold summits. Clouds are constantly falling by their weight, but as they melt away in the warm air below, by condensation they increase above.

The *nimbus* cloud is a dark-colored cloud from which rain is falling.

The *stratus* cloud is composed of broad, widely-extended cloud-belts, sometimes spread over the whole sky. It is the lowest cloud, and often rests on the earth. It is the night-cloud.

The *cumulus* cloud is made up of large cloud-masses looking like snow-capped mountains piled up along the horizon. It forms the summits of pillars of vapor, which, streaming up from the earth, are condensed in the upper air. It is the day-cloud. When of small size and seen only near mid-day, it is a sign of fair weather.

The *cirrus* (curl) cloud consists of light, fleecy clouds floating high in air. It is believed to be formed of spiculæ of ice or flakes of snow.

The *cirro-cumulus* is formed by small, distinct, rounded portions of the cirrus cloud, which separate from each other, leaving a clear sky between. Sailors call this a "mackerel sky." It accompanies warm, dry weather.

The *cirro-stratus* is produced when the cirrus cloud spreads out into long, slender strata. It forebodes storms.

The *cumulo-stratus* presents the peculiar forms called "thunder-heads." It is caused by a blending of the cumulus with the stratus, and is a precursor of thunder-storms.

RAIN is vapor condensed by the *sudden* cooling of the air in the upper regions. At a low temperature the vapor is frozen directly into *snow*. This may melt before it reaches the earth, and fall as rain. A sudden draught of cold air into a heated ball-room has sometimes produced a miniature snow-storm. The wonderful variety and beauty of snow-crystals are illustrated in the accompanying figure.

Fig. 179.

WINDS are produced by variations in the temperature of the air. The atmosphere at some point is expanded, rises, and colder air flows in to supply its place. This produces currents. The *land and sea breezes* of tropical islands are caused by the unequal specific heat of land and water. During the day the land becomes more highly heated than the water, and hence toward evening a sea-breeze sets in from the ocean. At night the land cools faster than the water, and so toward morning a land-breeze sets out from the land. *Trade-winds* are so named because by their regularity they favor commerce. A vessel on the Atlantic Ocean will sometimes, without shifting a sail, set steadily before this wind from Cape de Verde to the American coast. The air about the equator is highly heated, and, rising to the upper regions, flows off north and south. The cold air near the poles sets toward the equator to fill its place. If the earth were at rest this would make an upper warm current flowing from the equator, and a lower cold current flowing toward it. As the earth is revolving on its axis from west to east, the under current starting from the poles is constantly coming to a part moving faster than itself. It therefore lags behind. When it reaches the north equatorial regions it lags so much that it becomes a current from the northeast, and in the south equatorial regions a current from the southeast.

OCEANIC CURRENTS are produced in a similar manner. The water which is heated by the vertical sun of the tropics rises and flows toward the poles. The Gulf Stream, which issues from the Gulf of Mexico, carries the heat of the Caribbean Sea across the Northern Atlantic to the shores of Scotland and Norway. This tropical river flowing steadily through the cold water of the ocean, rescues England from the snows of Labrador. Should it by any chance break through the Isthmus of Panama, Great Britain would be condemned to eternal glaciers.

VARIOUS FORMS AND ADAPTATIONS OF WATER.—The great specific heat of water adapts it to exercise a marked influence on climate. Warm winds sweeping northward meet the colder air of the temperate regions and deposit their moisture. The latent heat which the vapor absorbed in the sunny South is set free, to temper the severity of our snowy climate. Thus, aërial and oceanic currents constitute a water circulation which is a natural steam apparatus on the grandest scale, since it has a boiler at the equator, and steam-pipes running over the entire globe. Water also equalizes the climate. It tends to prevent sudden changes of weather. In the summer it absorbs vast quantities of heat, which it gives off in the fall to moderate the approach of winter. In the spring the melting ice and snow drink in the warmth of the sunbeam, which else might prematurely coax forth the tender buds. Since so much heat is required to melt the ice and snow, they dissolve very slowly, and thus prevent in a measure the disastrous floods which would inevitably follow, if they passed quickly into the liquid state.

Water contains air, which is necessary for the support of fish. Just here occurs one of those happy coincidences which frequently startle the reverent searcher in Nature. Were water deprived of this air, it would be liable to explode at any moment when it happened to be heated much above $212°$. Every stove-boiler would need a thermometer. A teakettle would then require as careful

watching as now to attend a steam-engine, and our kitchens would witness frequent and most disastrous explosions. As it is, when the temperature rises above 212°, the extra heat passes off quietly and safely. Water expands with heat, like other liquids, and contracts, on cooling, down to 39° F. Then it slowly expands until it reaches 32° F., when it freezes. The bursting of water-pipes and pails is a familiar example of this exception. Under the operation of the general law, the water at the surface radiating its heat and becoming chilled, would contract and fall to the bottom, while the warm water below would rise to the top. This process would continue until the freezing-point was reached, when the whole mass would instantly solidify into ice. Our lakes and rivers would thus freeze solid every winter. This would be fatal to fish and aquatic vegetation. In the spring, the ice would not, as now, buoyant and light, float and melt in the direct sunbeam, but, lying at the bottom, would be protected by the non-conducting water above. The longest summer would not be sufficient to thaw the deeper bodies of water. Here we see another instance of prudent foresight. An exception is made to prevent these disastrous consequences.*
The cold water expands and rises to the top, thus protecting the warm water beneath, while ice itself,

---

* Certain metals—iron, bismuth, etc.—are also an exception to the general law. This fact adapts them for castings. Is not this equally a thoughtful provision for our wants?

being a non-conductor, preserves the temperature of the water quite uniform during the entire winter.

Water, in freezing, has a tendency to free itself from impurities. This furnishes a means of obtaining fresh water in Arctic regions. McClintock found that on each successive freezing the ice was purer, until, on the fourth time, he obtained drinking-water. If a barrel of vinegar freezes, we shall find the acid collected in a little mass at the centre of the ice.

When the dew collects at night sufficiently to form a covering upon the plants, being a non-conductor, it stops further radiation of heat. Thus, by a nice provision, the effect of radiation checks the radiation itself, as soon as the wants of the thirsty vegetation are supplied.

Water distills from the ocean and land as vapor, at one time cooling and refreshing the air, at another moderating its wintry rigor. It condenses into clouds, which shield the earth from the direct rays of the sun, and protect against excessive radiation. It falls as rain, cleansing the air and quickening vegetation with renewed life. It descends as snow, and, like a coverlet, wraps the grass and tender buds in its protecting embrace. It bubbles up in springs, invigorating us with cooling, healing draughts in the sickly heat of summer. It purifies our system, dissolves our food, and keeps our joints supple. It flows to the ocean, fertilizing the soil, and floating the products of industry and toil to the markets of the world. (See Chemistry, pp. 56-63.)

Practical Questions.—1. Why will one's hand, on a frosty morning, freeze to a metallic door-knob sooner than to one of porcelain? 2. Why does a piece of bread toasting curl up on the side exposed to the fire? 3. Why do double windows protect from the cold? 4. Why do furnace-men wear flannel shirts in summer to keep cool, and in winter to keep warm? 5. Why do we blow our hands to make them warm, and our soup to make it cool? 6. Why does snow protect the grass? 7. Why does water "boil away" more rapidly on some days than on others? 8. What causes the crackling sound in a stove when a fire is lighted? 9. Why is the tone of a piano higher in a cold room than in a warm one? 10. Ought an inkstand to have a large or a small mouth? 11. Why is there a space left between the ends of the rails on a railroad track? 12. Why is a person liable to take cold when his clothes are damp? 13. What is the theory of corn-popping? 14. Could vacuum-pans be employed in cooking? 15. Why does the air feel so chilly in the spring, when snow and ice are melting? 16. Why in freezing ice-cream do we put the ice in a wooden vessel and the cream in a tin one? 17. Why does the temperature generally moderate when snow falls? 18. What causes the singing of a teakettle. *Ans.* The escaping steam is thrown into vibration by the peculiar shape of the spout. 19. Why does sprinkling a floor with water cool the air? 20. How low a degree of temperature can be marked by a mercurial thermometer? 21. If the temperature be 70° F., what is it C.? 22. Will dew form on an iron bridge? On a plank walk? 23. Why will not corn pop when very dry? 24. The interior of the earth being a melted mass, why do we get the coldest water from a deep well? 25. Ought the bottom of a teakettle to be polished? 26. Which boils the sooner, milk or water? 27. Is it economy to keep our stoves highly polished? 28. If a thermometer be held in a running stream, will it indicate the same temperature that it would in a pailful of the same water? 29. Which makes the better "holder," woollen or cotton? 30. Which will give out the more heat, a plain stove or one with ornamental designs? 31. Does dew fall? 32. What causes the "sweating" of a pitcher? 33. Why is evaporation hastened in a vacuum? 34. Does stirring the ground around plants aid in the deposition of dew? 35. Why does the snow at the foot of a tree melt sooner than that in the open field? 36. Why is the opening in a chimney made to decrease in size from bottom to top? 37. Will tea keep hot longer in a bright or a dull teapot? 38. What causes the snapping of wood when laid on the fire? *Ans.* The expansion of the air in the cells of the wood. 39. Why is one's breath visible on a cold day? 40. What gives the blue color to air? *Ans.* The vapor it contains reflects the blue light of the sunbeam? 41. Why is light-colored clothing cooler in summer and warmer in winter than dark? 42. How does the heat at two feet from the fire compare with that at a distance of four feet? 43. Why does the frost remain later in the morning upon some objects than upon others? 44. Is it economy to use green wood? 45. Why does not green wood snap? 46. Why will a piece of metal dropped into a glass or porcelain dish of boiling water increase the ebullition? 47. Which can be ignited the more quickly with a burning-glass, black or white paper? 48. Why does the air feel colder on a windy day? 49. In what did the miracle of Gideon's fleece consist? 50. Could a burning lens be made of ice? 51. Why is an iceberg frequently enveloped by a fog? 52. Would dew gather more freely on a rusty stove than on a bright kettle? 53. Why is a clear night colder than a cloudy one? 54. Why is no dew formed on cloudy nights?

Electricity.

"That power which, like a potent spirit, guides
The sea-wide wanderers over distant tides,
Inspiring confidence where'er they roam,
By indicating still the pathway home;—
Through nature, quickened by the solar beam,
Invests each atom with a force supreme,
Directs the cavern'd crystal in its birth,
And frames the mightiest mountains of the earth;
Each leaf and flower by its strong law restrains,
And binds the monarch Man within its mystic chains."

HUNT

THALES, one of the seven wise men, knew that when amber is rubbed with silk it will attract light bodies, as straw, leaves, etc. This property was considered so marvellous that amber was supposed to possess a soul. From the Greek name of this substance (elektron) our word electricity is derived. The electrical force manifests itself in five different forms—(1) *Magnetic* electricity; (2) *Frictional* or statical electricity; (3) *Galvanic*, voltaic, or dynamic electricity; (4) *Thermal* electricity; (5) *Animal* electricity. These are intimately connected; their laws are strikingly related; they produce many effects in common; and each can give rise to the other.

MAGNETIC ELECTRICITY.

Magnetism treats of the properties of magnets.

A MAGNET is a body which has the power of attracting iron. The term is derived from the fact that an ore of iron possessing this property was first found at Magnesia, in Asia Minor. *Natural magnets* are generally known as lodestones (Saxon, *laedan*, to lead). The one worn by Sir Isaac Newton weighed only 3 grains, yet it was able to lift 746 grains, or nearly 250 times its weight. Their power does not increase in proportion to their size. One brought from Moscow to London weighed 125 lbs., but could support only about 200 lbs. The *artificial magnet* consists of a magnetized steel bar; if straight, it is called a *bar* magnet; if bent into the shape of the letter U, a *horse-shoe* magnet. A piece of soft iron, called the *armature*, is placed on the end.

The Poles.—If we insert a magnet in iron-filings, they will cling chiefly to its extremities, which are

Fig. 180.

termed the poles. The magnetic force will be exerted even through an intervening body. Lay a sheet of paper on a magnet and sprinkle iron-filings upon it. They will collect in curious lines, the *mag-*

Fig. 181.

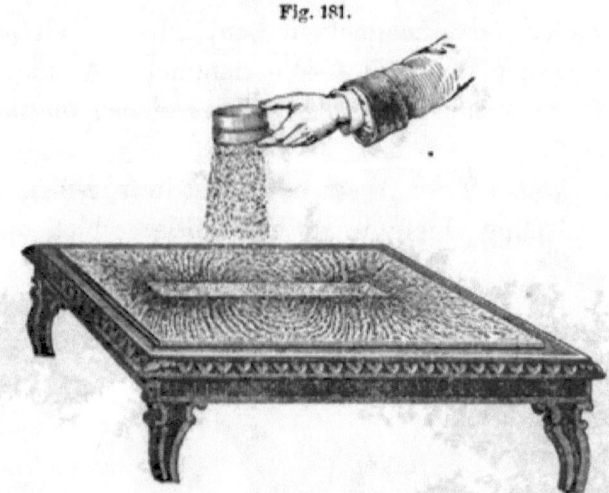

netic curves, radiating from the poles. If a small bar magnet be suspended so as to swing freely, one pole will point toward the north and the other toward

Fig. 182.

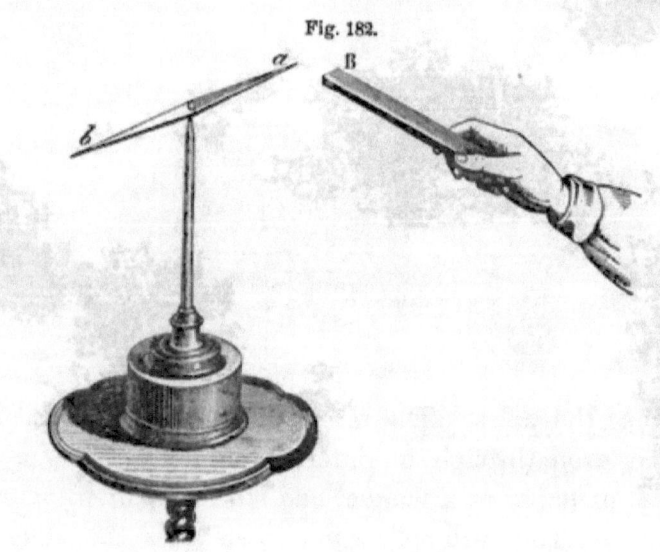

the south. The north pole of the magnet is called the positive (+), and the south pole, the negative (−). A magnet thus poised constitutes a *magnetic needle.* If we hold a magnet near a magnetic needle, we shall find that the south pole of one attracts the north pole, and repels the south pole of the other. This proves the law of magnetic attraction and repulsion—" *Like poles repel, and unlike poles attract.*"

MAGNETIC INDUCTION is the power a magnet possesses to develop magnetism in iron. If a piece of soft iron be brought near a magnet, it immediately assumes the magnetic state, which, however, it loses on being removed. In steel the change is permanent. The end of the bar next to the south pole of the magnet becomes the north pole of the new magnet, and *vice versa.* When opposite states are thus developed in the opposite ends of a body, it is said to be *polarized.* Whenever any object is attracted by a magnet, it is supposed first to be made a magnet (polarized) by induction, and then the attraction consists merely in that of unlike poles for each other. Thus we may suspend from a magnet a chain of rings held together by magnetic attraction.

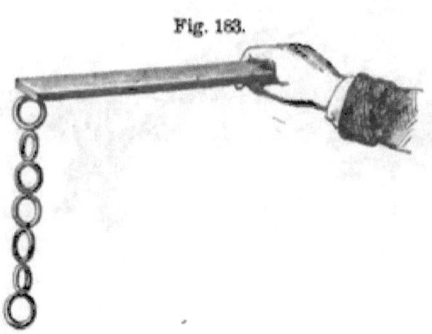

Fig. 183.

Each link is a magnet with its north and south poles. Each particle of the tuft of filings in Fig. 180 is a distinct, perfect magnet. A magnet does not lose any force by inducing magnetism. It rather gains strength by the reflex influence of the new magnet. An armature acts in this manner to strengthen a magnet. If we break a magnet even into the smallest fragments, each part will have a north and a south pole. It is explained by supposing that every molecule of iron contains two kinds of electric force which neutralize each other. When magnetized they are separated, but do not leave the molecule in which they reside. Each molecule is thus polarized, the two halves assuming opposite magnetic states, as shown in the figure. The light half of each little circle represents the positive, and the dark the negative side. All the molecules exert their negative force in one direction, and their positive in the other. The forces thus neutralize each other at the centre, but manifest themselves at the ends of the magnet.

Fig. 184.

HOW TO MAKE A MAGNET.—The following method is an excellent one. Place the inducing magnet on the unmagnetized one, as shown in the figure, and draw it from one end to the other several times, always carrying it back through the air in a circle to the starting-point.

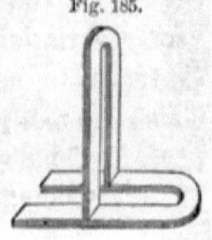

Fig. 185.

THE COMPASS is a magnetic needle used by mariners, hunters, surveyors, etc. It is very delicately poised over a card on which the "points of the compass" are marked. The needle does not often point directly N. and S. The *"line of no variation,"* as it is called, runs in an irregular course through

Fig. 186.

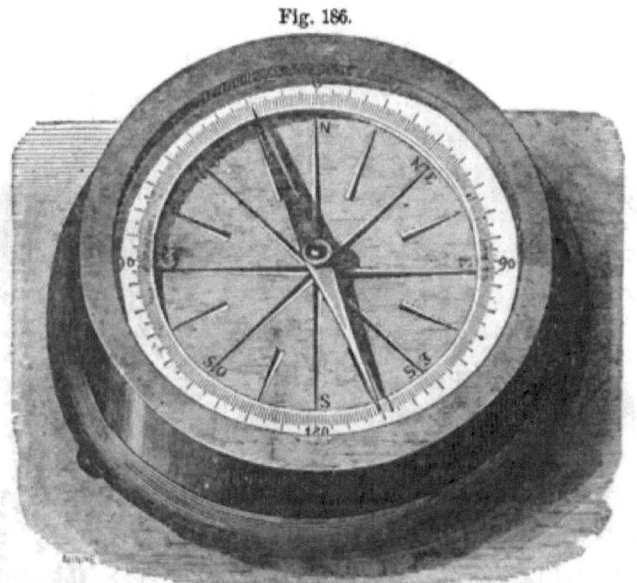

the United States from Cape Lookout across Lake Erie to Hudson's Bay. East of this, the variation (*declination*) is toward the west, and west it is toward the east. The needle is subject also to daily and yearly variations, as well as those which require centuries to complete. The needle is, however, "true to the pole," although it shifts thus every hour in the day. It does so only in obedience to the laws which control its action.

WHY THE NEEDLE POINTS NORTH AND SOUTH.—The earth is a great magnet. This gives direction to the needle. Variations which are constantly taking place in the terrestrial magnetism produce corresponding changes in the needle. Suppose a magnet N S passing through the centre of a small globe. The needle *s n* will hang parallel to it, as in Fig. 187, its north pole being attracted by the south pole of the magnet, and *vice versa*. If the globe be turned, Fig. 188, the north pole of the needle will bend—*dip*, as it is termed—downward. If the globe be turned in the other direction, the south pole of the needle will dip in the same manner. Similar phenomena are noticed in the compass. At the equator it is horizontal, but *dips* whenever taken north or south. An unmagnetized needle, if carefully poised, in our latitude, on being magnetized, immediately settles down, as if the north end were the heavier. This difficulty is remedied by making the north end of the needle lighter, and also by suspending a little weight upon the south end. The reverse is true in the southern hemisphere.

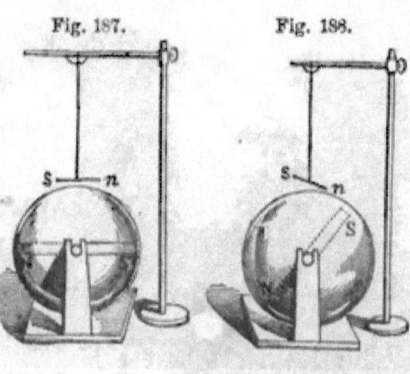

A dipping-needle is poised as shown in Fig. 189.

At the equator it hangs horizontally, but declines as

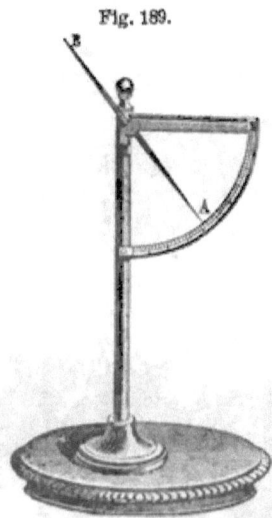

Fig. 189.

it is carried north, until, at a place near Hudson's Bay, as discovered by Captain Ross in 1832, it becomes vertical. This point is called the North magnetic pole. Strangely enough, it does not coincide with the geographical pole. The South magnetic pole has not yet been found. From the experiments we have made, we see that the end of the needle which points toward the N. pole of the earth, is really its S. pole.

THE EARTH INDUCES MAGNETISM.—All iron bars standing vertically (which in this latitude is not far from the line of the dip) possess slight magnetic properties. The upright parts of an iron fence, lightning rods, standards of chairs and desks, etc., on being tested by the magnetic needle, will be found to possess north polarity in the end next the ground, and south polarity in the other. The polarity of the lodestone has doubtless been caused in this manner in the lapse of ages.

FRICTIONAL ELECTRICITY.*

This is electricity developed by friction. One's hair often crackles under a gutta-percha comb. A cat's back, when rubbed in a dark room, emits sparks. In cold, frosty weather, a person, by shuffling about in his stocking-feet upon the carpet, can develop so much electricity in his body that he can ignite a jet of gas by simply applying his finger to it.

THE ELECTROSCOPE is an instrument for detecting the presence of electricity. Bend a glass tube, and suspend from it a couple of pith-balls, as shown in the figure. Two strips of gold-leaf, hung in a glass jar (Fig. 191), form a more delicate test. This instrument is so sensitive, that a slight flap of a silk handkerchief against the cover will cause the leaves to diverge.

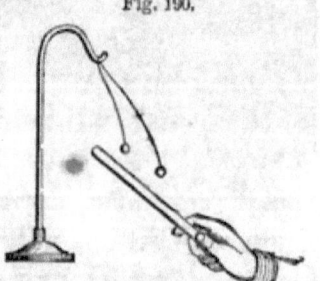

Fig. 190.

TWO KINDS OF ELECTRICITY.—If a warm, dry glass tube (a lamp-chimney will answer) be rubbed with a silk handkerchief, a crackling sound is heard. If the tube be held near the face, we shall experience a sensation like that given by a cobweb. The tube

---

* The term *static* is applied to frictional electricity, and *dynamic* to galvanic. The former indicates a force at rest; the latter, one in motion.

will attract bits of paper, straw, feathers, etc. Present it to the pith-balls in the electroscope (Fig. 190). They will be attracted for an instant, and will then fly from the tube and from each other, apparently in the utmost disgust. Electrify a stick of sealing-wax and present it to the balls. They will act in the same manner. If we touch one ball to the excited glass, and the other to the excited wax, they will not, as before, fly from each other, but will rush together at once. Present the glass to a ball: it will fly off when electrified. Present the glass again, and it will be repelled. Substitute the wax, and it will be attracted. Offer now the glass, and it will eagerly bound toward what it just before spurned. If the glass be held on one side of a ball and the wax on the other, it will fly between the two, carrying the electricity back and forth. From this we conclude (1), that there are two kinds of frictional as of magnetic

Fig. 191.

electricity; and (2), *like electricities repel each other, and unlike attract.* The electricity from the glass is termed vitreous or positive [+], and that from the wax, resinous or negative [−].

In the following list, each substance becomes positively electrified when rubbed with the body following it; but negatively, with the one preceding it. (*Ganot.*)

1. Cat's fur.
2. Flannel.
3. Ivory.
4. Glass.
5. Cotton.
6. Silk.
7. The hand.
8. Wood.
9. Shellac.
10. Resin.
11. The metals.
12. Sulphur.
13. Caoutchouc.
14. Gutta-percha.
15. Gun-cotton.

THEORY OF ELECTRICITY.—Of the nature of electricity we know little. The positive and negative forces exist in every body in a state of equilibrium. When this is disturbed by friction, chemical action, etc., both are set free. We cannot develop one without the other. The opposite kinds manifest themselves at opposite parts of the surface, as in a magnet; it is therefore called a *polar force*. The slightest causes disturb the electric equilibrium. "In cutting a slice of meat, there may pass between the steel knife and silver fork enough electricity to move the needle of a telegraph." Yet the delicate balance of the opposing forces is so soon readjusted that we are unconscious of the change.

CONDUCTORS AND INSULATORS.—A body which allows the electric force to pass freely through it is termed a *conductor;* one which does not, is called a *non-conductor*, or *insulator*. Copper is one of the best conductors, and hence it is used in all electrical experiments. Glass is one of the best insulators. A body is said to be insulated when it is supported by some non-conducting substance, usually glass. The air, when dry, is a non-conductor, but when moist becomes a good conductor. Hence, the electricity can be retained on an insulated body in a dry atmosphere, but is soon dissipated in a damp one. Electricity can be collected only by means of insulation. It can be developed by rubbing an iron rod, but is lost as fast as formed, by passing off through the metal to the hand. A glass rod does not conduct it to the body, so it is retained until it gradually dissipates in the air. The following list contains the most common conductors and insulators.

Best Conductors.

Metals.	Vegetables.	Ice.	Glass.
Charcoal.	Animals.	Dry Wood.	Wax.
Flame.	Linen.	Caoutchouc.	Sulphur.
Minerals.	Cotton.	Dry Paper.	Amber.
Water.	Dry Wood.	Air.	Shellac.
Iron.	Ice.	Silk.	*Best Insulators.*

THE ELECTRICAL MACHINE consists (1) of a glass *wheel* turned by a crank; (2) of a pair of *rubbers* covered with leather and spread with an amalgam (a mixture of tin, zinc, and mercury) which hastens the development of electricity; (3) of a *comb* or fork with fine points, since pointed bodies always favor the reception or dispersion of electricity; (4) of a *prime conductor*—a brass cylinder insulated by a **glass standard** so that the electricity cannot pass to the ground, and rounded at the ends so that it may not escape too rapidly into the atmosphere.

At the commencement, the whole apparatus is in a state of equilibrium. By the friction, positive electricity is developed on the glass, and negative on the

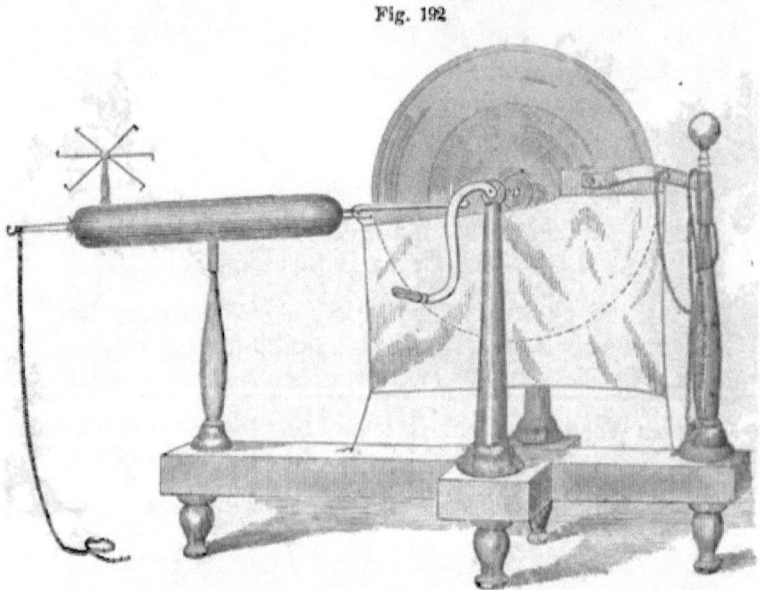

Fig. 192

rubber. The negative escapes along the chain to the ground—the common reservoir. The positive, kept on the glass by the silk flaps, is carried around to the points. Here it attracts the negative electricity of the prime conductor, and the two forces, clashing together, form tiny sparks. The positive electricity naturally present in the prime conductor is thus left insulated, and the prime conductor is said to be *charged* with positive electricity. If the negative conductor be insulated, the rubber will soon become charged with negative electricity, and the action of the machine will nearly cease. If the air be dry, the rubber freshly spread with amalgam, and the glass well rubbed with warm flannel, a sharp crackling noise will be heard, flashes will follow the wheel around, while sparks can be obtained from the prime conductor at a distance of several inches. The pith-ball electroscope, when charged and repelled by the prime conductor, will be quickly attracted by the rubber. This indicates the opposite electricities in them.

INDUCTION.—The influence of an electrified body over other bodies near it is termed *electrical induction*.

Fig. 193.

Thus, let a small insulated conductor be placed near the ball at the end of the prime conductor of an electrical machine. On charging the prime conductor the motion of the pith-balls will show that the small conductor has also become charged. On testing with the electroscope, we shall find that the end next the positive prime conductor is negative, and the other end positive. As opposite electricities are thus developed at the opposite extremities of the conductor, it is *polarized*. Place several conductors, as shown in Fig. 194, connecting

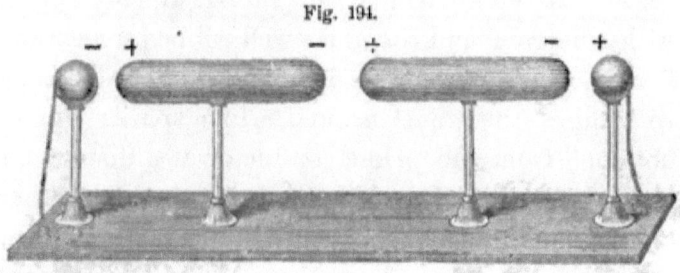

Fig. 194.

the copper ball at the right with the positive pole, and the one at the left with the negative pole of the electrical machine. The conductors will be charged and polarized by induction.

Faraday's theory of induction assumes (1) that the electricity acts between the different molecules of a body, as between the different conductors in the last experiment—that each molecule becomes polarized, and in turn polarizes its neighbors, and that thus at last every molecule has opposite electricities on its opposite sides; (2) that the molecules of non-conductors become polarized and *retain* their electricities, while the molecules of conductors become polarized and *discharge* their electricities into the adjacent molecules. The positive force thus passing from one molecule to another of a conductor accumulates at one end, and the negative, moving in the opposite direction, collects at the other end. Let P (Fig. 195) represent the end of the positive conductor and N that of the small conductor in Fig. 193; and let the small circles represent molecules of air lying between the two—the lighter half indicating the positive and the darker half the negative side. The molecules of air being non-conducting, on being polarized from the influence of P, the prime conductor, retain their electricities, but polarize each other in succession until N is reached. This being a conducting body, its molecules impart their electricity from one to the other, until the negative electricity collects at one end and the positive at the other.

ATTRACTION AND REPULSION.—Every case of attraction or repulsion is preceded by induction. "The

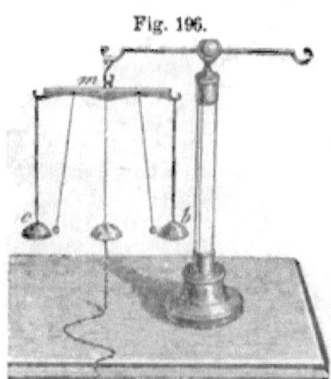

Fig. 196.

electric chime" illustrates this very prettily. It consists of three bells, two of which, c and b, are hung by brass chains, while the middle one is insulated above by a silk cord, and connected below with the earth by a chain. The balls between them are also insulated. The outer bells becoming charged with positive electricity from the prime conductor, polarize the balls by induction through the intervening air. The balls being then attracted to the bells, are charged and immediately repelled. Swinging away, they strike against the middle bell, discharge their electrical force, and are forthwith attracted again. Flying to and fro, they ring out a merry, electrical song. The *dancing image* is another illustration. It consists of a little pith-ball figure placed between two metallic plates, the upper one hanging from the prime conductor, and the lower one connected with the earth. The dance is conducted in a remarkably lively manner by alternate attraction and repulsion.*

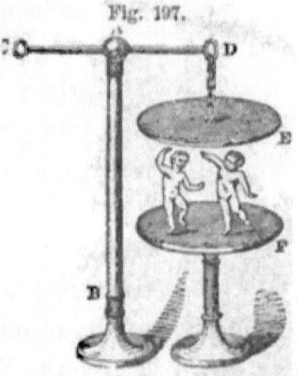

Fig. 197.

THE LEYDEN JAR consists of a glass jar coated inside and outside, nearly to the top, with tinfoil. It is fitted with a cover of baked wood, through which passes a wire with a knob at the top, and below, a chain extending to the inner coating. The jar is *charged* by bringing the knob near the prime conductor of the electrical machine, while the outer coating communicates freely with the earth. Bright sparks will then leap in rapid succession to the inner coating, while simi-

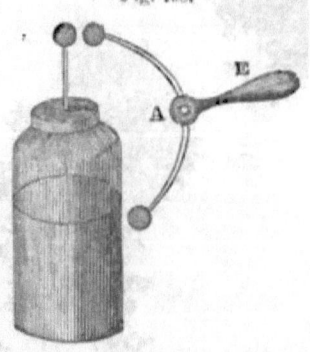

Fig. 198.

---

* A very slow motion should be given to the electrical wheel, and a pin thrust into the heel of the image will add much to the stamp of the tiny feet.

lar ones will pass off from the outer coating. The jar is *discharged* by holding one knob of the "discharger" E, upon the outer coating, and the other upon the knob of the jar. The equilibrium will be restored with a sharp snap and a brilliant flash. Minute particles are detached from the solid conductors, and, burning, give color and brilliancy to the spark.*

Explanation.—The charging of a jar with electricity is entirely different from the process of filling one with water. The glass can as well be in the form of a pane. The only essentials are *two conducting surfaces separated by a non-conducting body*. The tinfoil acts only as a conductor to convey the electricity. This is finely illustrated by the "Leyden jar with movable coatings," which may be charged and then taken apart. Very little electricity can be obtained from the glass, either of the tin coatings, or any two of the parts combined. When put together again, the jar can be discharged in the usual

---

* Professor Muschenbroek, of Leyden, discovered the principle of the Leyden jar in the following curious way. While experimenting, he held a bottle of water to the prime conductor of his electrical-machine. Noticing nothing peculiar, he attempted to investigate its condition. Holding the bottle with one hand, he happened to touch the water with the other, when he received a shock so unexpected, and so unlike anything he had ever felt before, that he was filled with astonishment. It was two days before he recovered from his fright. A few days afterward, in a letter to a friend, the Professor innocently remarked, that he would not take another shock for the whole kingdom of France.

manner. Fig. 199 represents an enlarged section of the side of a Leyden jar: 1 indicates the inner coating; 2, the outer coating, and the circles between, the molecules of glass. The sparks of positive electricity from the prime conductor are distributed by the inner coating, over the jar. The molecules of glass are polarized, while the outer coating becomes charged by induction with negative electricity. A quantity of positive electricity corresponding to the positive received by the inner coating escapes from the outer coating. If the jar be insulated so that this is unable to leave, the passage of the sparks will soon cease. If a finger be held near the outer coating, a spark will leap to it every time one enters the jar. The jar, therefore, when charged, contains no more electricity than in its natural state. It is only differently distributed.

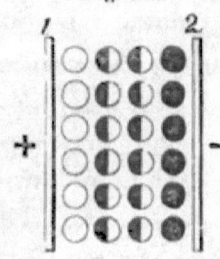

Fig. 199.

THE ELECTRICITY IS ON THE SURFACE.—Each molecule within the surface of a solid, insulated conductor gives up its electricity with equal freedom in every direction; therefore it cannot become charged. Each molecule on the surface, however, receiving electricity from the particles behind it, and having non-conducting particles of air before it, must become charged. A bomb-shell can therefore hold as much electricity as a cannon-ball.

THE EFFECT OF POINTS.—A pointed wire held near

the prime conductor will quietly draw off all its electricity, which will be seen apparently clinging to the point like a little glowing star. If we fasten a pointed wire to the prime conductor, it will discharge the electricity in a *brush* of flame, silently, but so rapidly that even the pith-balls will not reveal its presence in the conductor. If we hold one cheek near the point, we shall feel a current of air setting away from it. This is strong enough to deflect the flame of a candle. The particles of air near the point become polarized, are attracted, give up their negative electricity, and, being charged with positive electricity, are repelled; new ones take their place, and thus a current is established. The *electric whirl*, mounted on the prime conductor (Fig. 192), illustrates this action. As each molecule of air is repelled from a point, it reacts with equal force against the point. This is sufficient to set the light wire-wheel in rapid rotation.

ATMOSPHERIC ELECTRICITY.—If, with the friction upon a small glass wheel, so much electricity is developed, what immense quantities must be produced by moving masses of air, clouds, etc.! Added to this, are the effects of heat, chemical changes, and the varied processes of nature—all of which disturb the electrical equilibrium. The air, therefore, is almost constantly electrified. In clear weather it is in a positive state, but in foul weather it changes rapidly from positive to negative, and *vice versa*. Dr. Livingstone tells us that in South Africa the hot wind which blows over the desert is so highly electrified, that a bunch of ostrich feathers held for a few seconds against it becomes as strongly charged as if attached to an electrical-machine, and will clasp the hand with a sharp, crackling sound.

LIGHTNING is only the discharge of a Leyden jar on the grand scale upon which Nature performs her operations.* Two clouds charged with opposite electricities, and separated by the non-conducting air, approach each other. When the tension becomes sufficient to overcome the resistance, the two forces rush together with a blinding flash and terrific peal. The lightning moves along the line where there is the least resistance, and so describes a zigzag course. If we can trace the entire length, we call it *chain-lightning;* if we only see the flash through intervening clouds, it is *sheet-lightning;* and if it is the reflection of distant discharges, we term it *heat-lightning.* The report is caused by the clashing of the atoms of displaced air. The rolling of the thunder is produced by the reflection of the

---

* The identity of lightning and frictional electricity was discovered by Franklin. He made a kite of a silk handkerchief, and fixed at the top a pointed wire. He elevated this during a thunder-storm, tying at the end of the hemp string a key, and then insulating the whole by fastening it to a post with a long piece of silk lace. On presenting his knuckles to the key, he obtained a spark. So great was his joy, that he is said to have burst into tears. He afterward charged a Leyden jar, and performed other electrical experiments in this way. These attempts were attended with very great danger. A few years after, Prof. Richman drew in this manner from the clouds a ball of blue fire as large as a man's fist. It struck him lifeless.

sound from distant clouds. Sometimes the clouds and the earth become charged with opposite electricities, separated by the non-conducting air. The spark from the discharge of this huge Leyden jar is a bolt that often causes fearful destruction.

Fig. 200.

The Aurora Borealis—"Northern lights"—is probably caused by the passage of electricity through the rarefied atmosphere of the upper regions. It may be beautifully imitated, on a small scale, by passing a succession of sparks from the prime conductor through a long glass tube from which the air is nearly exhausted. The intimate relation between the aurora and magnetism is shown from the fact that the magnetic needle is disturbed when the aurora is visible, and seems to tremble as the streamers dart to and fro. The telegraph is frequently worked by the current of electricity which passes along the wire on these occasions, thus for a time dispensing with the line-batteries. *Geissler's Tubes* are filled with rarefied gases, and then sealed. **When a current** of electricity is passed through them, **the** richest tints and variegated bands of color are exhibited. *Gassiot's Cascade* consists of a glass goblet coated with tinfoil on the inside. This is placed on the air-pump, and covered with a receiver which has a sliding-rod passing through the **top. The air is then** exhausted, and the rod brought into contact with the tinfoil. One conductor of the electrical-machine is connected with the rod, and the other with the pump-plate. The electricity will flow over the sides of the cup in a shower of soft undulations and delicate blue light.

Lightning-rods were invented by Franklin. They are based on the principle that electricity always seeks the best conductor. The rod should be pointed at **the top with** some metal which will not easily corrode. If constructed in separate parts, they should be securely jointed. The lower end should extend into water, or else deep into the damp ground, beyond a possibility of any drought rendering the **earth about it a** non-conductor, and be packed about **with ashes or** charcoal. If the rod is of iron, it needs to be much larger than if of copper, which is a better conductor. Every elevated portion of the building should be protected by a separate rod. Chimneys in which fire is constantly kept need especial care, because of the ascending column of vapor and smoke. Water conductors, tin roofs, etc., should be connected with the damp ground or the lightning-rod, that they may aid in conveying off the electricity. The value of a lightning-rod consists, most of all, in its power of quietly restoring the equilibrium between the earth and the clouds. By erecting lightning-rods, we thus lessen the liabilities of a sudden discharge. Providence has provided largely against this catastrophe. "God has made a harmless conductor in every leaf, spire of grass, and twig. A common blade of grass, pointed by Nature's exquisite workmanship, is three times more effectual than the finest cambric needle, and a single pointed twig than the metallic point of the best-constructed rod." Every drop of rain, and every snow-flake, falls charged with the electric force, and thus quietly disarms the clouds of their terror. The balls of electric light, called by sailors "*St. Elmo's fire*," which sometimes cling to the masts and shrouds of vessels, and the flames seen to play about the points of bayonets, indicate the quiet escape of the electric force from the earth toward the clouds.

VELOCITY OF ELECTRICITY.—The duration of the flash has been estimated at one-millionth of a second. Some idea of its instantaneousness can be formed from the fact that the spokes of a wheel revolved so rapidly as to become invisible by daylight can be distinctly seen by the spark from a Leyden jar. The trees swept by the tempest, when seen by a flash of lightning, seem motionless, while a cannon-ball, in swift flight, appears poised in mid-air. Wheatstone considered the velocity of lightning through a copper wire to be 288,000 miles per second.

EFFECTS OF FRICTIONAL ELECTRICITY.—I. *Physical.* —Discharges from a large battery will melt rods of the various metals, perforate glass, split wood, magnetize steel bars, etc. 1. Let a person stand upon an insulated stool and become charged from the prime conductor. His hair, through repulsion, will stand erect in a most ludicrous manner. On presenting his hand to a little ether contained in

Fig. 201.

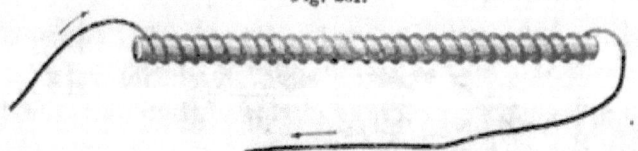

a spoon, a spark leaping from his extended finger will ignite it. If he hold in his hand an icicle, the spark will readily dart from it to the liquid. 2. A card held between the knob of a Leyden jar and that of the discharger, will be punctured by the spark. 3. A piece of steel may be magnetized by the discharge from an ordinary Leyden jar. Wind a covered copper wire around a steel bar, as in Fig. 201, or simply enclose a needle in a small glass tube around which the wire may be wound. On passing the spark through the wire, the needle will attract iron-filings. 4. When strips of tinfoil are

Fig. 202.

pasted on glass, and figures of various curious patterns cut from them, the electric spark leaping from one to the other presents a beautiful appearance. The *diamond Leyden jar* and the *spiral tube* illustrate these effects in a brilliant manner. 5. If a battery be discharged through a wire too small to conduct the spark, the electricity is changed to heat, and if sufficiently small, the wire will be fused into globules or dissipated in smoke.

The fact that the electric force is thus converted into vibrations of heat and light, would seem to indicate that, like them, it is only a mode of motion.

II. *Chemical effects.*—The "electric gun" is filled with a mixture of oxygen and hydrogen gases. A spark causes them to combine with a loud explosion and form water. The sulphurous smell which accompanies the working of an electrical-machine, and is noticed in places struck by lightning, is owing to the production of ozone, a peculiar form of the oxygen of the air. (See Chemistry, p. 38.)

III. *Physiological effects.*—A very slight charge from a Leyden jar produces a contraction of the muscles and a spasmodic sensation in the wrist. A stronger one affects the body, and becomes painful and even dangerous. The shock may be given to a large number of persons simultaneously by joining hands. The Abbé Nollet once shocked in this way a regiment of 1,500 soldiers.

GALVANIC ELECTRICITY.

Galvanic or Voltaic electricity is produced by chemical action.* These names are given in honor of the two Italian philosophers who made the first discoveries in this branch of electricity.

GALVANI'S DISCOVERY.—In the year 1790 Galvani was engaged in some experiments on animal electricity. For this purpose he used frogs' legs as electroscopes. He had hung several of these upon *copper* hooks from the *iron* railing of the balcony, in order to see what effect the atmospheric electricity might have upon them. He noticed, to his surprise, that when the wind blew them against the iron supports, the legs were convulsed as if in pain. After repeated experiments, Galvani concluded that this effect was produced by what he termed animal electricity, that this electricity was different from that caused by friction, and that he had discovered the agent by which the will controls the muscles.

---

* The pupil, on recalling the definition of Natural Philosophy, will readily perceive that galvanic electricity is a connecting link between philosophy and chemistry. Its cause is chemical, while its effects are both chemical and philosophical. It is oftentimes ranked as a part of what is termed Chemical Physics.

VOLTA'S DISCOVERY.—Volta rejected the idea of animal electricity, and after 27 years of incessant study, discovered that the frog was not the source of the electricity, but "only a moist conductor, and was not as good as a wet rag for that purpose." He applied this view to the construction of "Volta's pile." This is composed of plates of zinc and copper, between which are laid pieces of flannel moistened with an acid or saline solution. We can easily form a simple voltaic pile by placing a silver coin between our teeth and upper lip, and a piece of zinc under our tongue. On pressing the edges of the two metals together, we shall perceive a peculiar taste, while a flash of light will pass before the closed eyes. Volta believed that the contact of two dissimilar metals develops electricity. His theory has given place to the chemical one which we shall now notice.

THE SIMPLE GALVANIC CIRCUIT.—If we place a strip of zinc in a cup of water well acidulated with sulphuric acid (oil of vitriol), a chemical action will at once commence. Little bubbles of hydrogen gas will gather on the metal, while the zinc rapidly dissolves. If we now immerse the zinc in mercury, the surface will become as bright as a mirror. Replace the strip in the

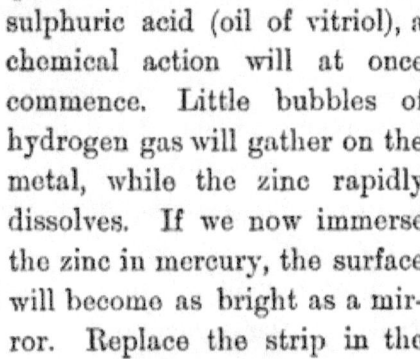
Fig. 204.

cup, and the acid will have no effect upon it. The reason of this action is not understood, but all zinc used in galvanic batteries is thoroughly and frequently amalgamated in this manner. Now put a strip of copper in the acid. As long as the two metals remain separate no change takes place, but as soon as they touch, or are connected by wires as in the figure, chemical action begins, and the bubbles of hydrogen gather upon the *copper* instead of the *zinc* as before. The copper will not be changed, but the zinc will waste away. As soon as the wires are separated the action ceases, and, in the dark, a minute spark is seen.

The ends of the wires are termed *poles* or *electrodes*. The copper pole is positive and the zinc negative. (These names may be easily remembered if we associate the p's with copper and positive, and the n's with zinc and negative.) Platinum strips are often fastened to the ends of the wires to act as *electrodes*, in order to withstand the corrosive liquids in which we may wish to place the poles. The joining of the wires is termed *closing the circuit*, and separating them *breaking the circuit*. Two metallic plates combined in this manner form a *voltaic pair*. The two metals must be dissimilar (one positive and the other negative), and must be immersed in a liquid which is capable of producing a chemical effect on only one of them. If both are equally acted upon, no current will be established, since the electricity set free by each will neutralize that developed by the other. The metal which sets free the electricity is termed the positive, and the other the negative plate.*

The chemical change which takes place in the voltaic pair may be very *simply* explained as follows: Each molecule of water is composed of an atom of oxygen and two of hydrogen; the former unites with the zinc, forming oxide of zinc. The sulphuric acid combines with this, making sulphate of zinc, which dissolves in the water. The hydrogen being set free rises to the surface and escapes. (For the replacement theory, see Chemistry, p. 51.)

Why the hydrogen comes off from the copper plate.— For simplicity of illustration, we shall suppose a row of water molecules† extending from the zinc to the copper plate. The negative oxygen of the molecule of water nearest the positive zinc is attracted to that plate, while the positive hydrogen is repelled. The atom thus driven off seeks refuge with the oxygen of the next molecule, and dispossesses its hydrogen. This atom in turn robs the third molecule of its oxygen, and so on until the last

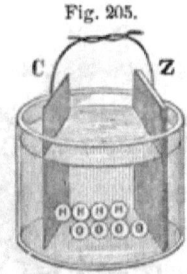

Fig. 205.

---

* It should be noticed that the terms are reversed when applied to the plates and the poles. The zinc pole is negative, but the zinc plate is positive; the copper pole is positive, but the copper plate is negative. We thus see that the plates when placed in the liquid become polarized, as is represented in the figure.

† In figure 205, a molecule of water is represented, for convenience, as consisting of only one atom of hydrogen and one of oxygen.

molecule is reached, when the atom of hydrogen, attracted by the negative copper, gives up to it its positive electricity, and then flies off into the air. Each atom of escaping hydrogen imparting its electrical force, adds to the current of electricity.

The Voltaic Current.—The word "current" is frequently used in electricity, but should not be understood to indicate the passage of a fluid, like the flow of water in a stream, but a mere transmission of the electrical force. Thus, if a long pipe were perfectly filled with water, a drop added at one end would thrust out a corresponding one at the other, which would not, however, be the identical one dropped in, since the force alone would traverse the length of the pipe. In the voltaic pair the current of positive electricity sets out from the positive zinc through the liquid to the negative copper, thence through the wire around again to the zinc. If the circuit is broken, the current manifests itself at the copper pole. There is also a negative current passing in the opposite direction; but, to avoid confusion, whenever the term current is used, the positive is intended.

In galvanic as in frictional electricity, when the current passes through a conducting substance, as a wire, rod, etc., the force is transmitted, not on the surface, as is sometimes said, but through the *entire thickness of the body.* Each molecule, becoming polarized and charged, discharges its force into the next molecule, and so on. The current thus moves by a rapid succession of polarizations and discharges of the molecules of the conductor. With what inconceivable rapidity must these successive changes take place in an iron wire to transmit the electric force, as in recent experiments, from San Francisco to Boston and return in one minute!

A BATTERY consists of several voltaic pairs so combined as to increase the strength and steadiness of the electric current.

SMEE'S BATTERY.—Each cell consists of two plates of zinc and one of silver suspended between them. They are clamped together with screws and hung in a glass jar filled with dilute sulphuric acid. Since bubbles of hydrogen gas tend to collect on the smooth surface of the silver and interrupt the action, it is roughened with finely divided platinum.

Fig. 206.

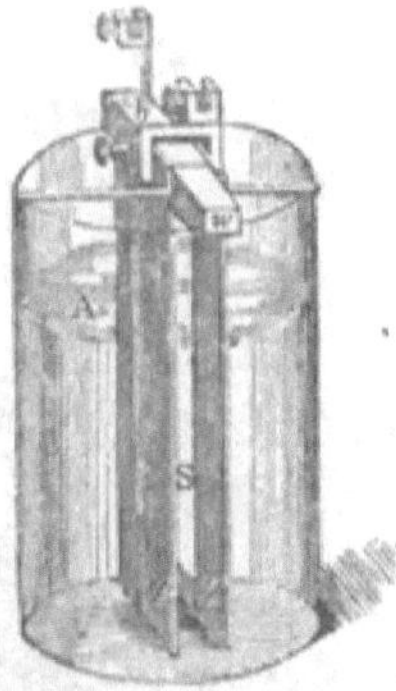

GROVE'S BATTERY is what is termed a "two-fluid battery." The outer glass jar contains dilute sulphuric acid, in which is placed a hollow zinc cylinder with a slit at the side to allow a free circulation of the liquid. The inner cup is of porous earthenware, and is filled with strong nitric acid (aqua fortis), in which is suspended a thin strip of platinum.

Fig. 207.

Chemical change.—The water in the outer cup is decomposed, the oxygen uniting with the zinc and the sulphuric acid with both, to make sulphate of zinc. The hydrogen, however, does not escape, as in Smee's battery, but passes into the inner cup and tears apart the nitric acid, forming water and nitric oxide. The latter is at first absorbed by the liquid, but soon begins to escape in corrosive, blood-red fumes. If the zinc is properly amalgamated, no action will take place while the poles are separated, and the battery will remain quiescent, like a sleeping giant, but the instant the wires are connected the liquid will begin to boil with the evolution of the gas, while the electric force will leap to the poles. (Rev. Chem., p. 47.)

Advantages of this Battery.—(1.) The hydrogen does not collect on the negative (platinum) plate, since it is absorbed by the nitric acid. (2.) The liquid formed in the inner cup is an excellent conductor of electricity. (3.) Platinum is a more perfect negative metal than copper, since it is not acted upon by the acid, and thus does not tend to start a counter-current; therefore platinum and zinc make a better voltaic pair than copper and zinc. (4.) The additional decomposition of the nitric acid sets free a great quantity of electricity.

BUNSEN'S BATTERY (Fig. 209) differs from Grove's merely in substituting a carbon rod for the platinum strip in the inner cup. This, being an excellent conductor, answers the same purpose and is much cheaper.

DANIELL'S CONSTANT BATTERY has an outer copper

cup holding a solution of blue vitriol, and an inner porous cup containing a zinc rod and dilute sulphuric acid. *The sulphate of copper battery* consists of a large zinc cylinder suspended in a copper jar containing a solution of sulphate of copper (blue vitriol).

QUANTITY AND INTENSITY.—A battery may develop a great quantity of electricity having a low degree of intensity, or a small quantity having a high intensity. Thus a cup of boiling water is intensely hot, while a hogshead full of that which is only blood-warm contains a great quantity of heat. The intensity of the electric force depends on the *number* of cells; the quantity, on their *size*. An intensity battery is formed by joining the zinc plate of one cell to the copper of the next; a quantity battery, by joining all the zinc and all the copper plates together. The former method is preferable when great resistance is to be overcome.

COMPARISON OF FRICTIONAL WITH GALVANIC ELECTRICITY.—Frictional electricity is noisy, sudden, and convulsive; galvanic is silent, constant, and powerful. The one is a quick, violent blow; the other a steady, uniform pressure. Intensity is the characteristic of the former, quantity of the latter. The lightning will leap through miles of intervening atmosphere, while the galvanic current will follow a conductor around the globe, rather than jump across the gulf of a half inch of air. The most powerful frictional machine would be insufficient for telegraphing; while despatches have been sent across the ocean with a tiny battery composed of "a gun-cap and a strip of zinc, excited by a drop of water the bulk of a tear." To decompose a grain of water would require over 6,000,000 discharges from a Leyden jar—enough electricity to charge a thunder-cloud 35 acres in area; yet a few galvanic cups would tear apart that amount of water in perfect ease and silence. Faraday immersed a voltaic pair, composed of a wire of platinum and one of zinc, in a solution of 4 oz. of water and one drop of oil of vitriol. In three seconds this produced as great a deviation of the galvanometer needle (Fig. 212) as was obtained by 30 turns of a powerful plate-glass machine. If this had been concentrated in one-millionth of a second, the duration of an electric spark, it would have been sufficient to kill a cat; yet it would require 800,000 such discharges to decompose a grain of water.

I. PHYSICAL EFFECTS OF VOLTAIC ELECTRICITY.—1. *Heat.*—If a current of electricity be passed through a wire too small to conduct it readily, it is converted into heat. The poorer the conducting power of the wire, and hence the greater the resistance, the more readily the change is produced. With 10 or 12 of Grove's cups several inches of fine steel wire may be thus fused, or even dissipated into smoke; and with a powerful battery, several yards of platinum wire (the poorest conductor) may be made glowing hot. Torpedoes and blasts are fired on this principle. Two copper wires leading from the battery to the spot are separated in the powder by a short piece of small steel wire. When the circuit is completed, the fine wire becomes red hot and explodes the charge.

2. *Light.*—In closing or breaking the circuit we produce a spark, the size of which depends on the intensity of the current. With several cells, beautifully variegated sparks are obtained by fastening one pole to a file and rubbing the other upon it. When charcoal or gas-carbon electrodes are used with a powerful battery, on slightly separating the points, the intervening space will be spanned by an arch of the most dazzling light. The flame, reaching out from the positive pole like a tongue, vibrates around the negative pole, licking now on this side and now on that. The heat is most intense. Platinum melts in it like wax in the flame of a candle, the metals burn with their characteristic colors, and even lime, quartz, etc., are fused. The effect is not produced by *burning* the charcoal points, since in a vacuum it is equally bril-

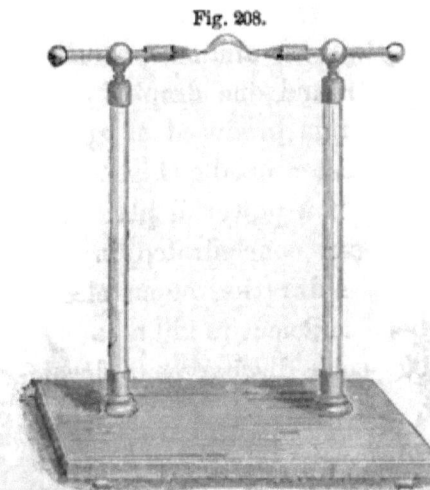

Fig. 208.

liant. The cost of the electric light and the intensity of the illumination, which renders the shadows extremely dense, have prevented its general use. It is interesting to notice that in the battery there is zinc burning, *i. e.*, combining with oxygen, but without light or heat; in the electric light the real force of the combustion is revealed. We may thus transfer the light and heat to a great distance from the fire.

II. CHEMICAL EFFECTS.—1. *Decomposition of Water.*—If the platinum electrodes are held a little distance apart in a cup of water, little trains of tiny

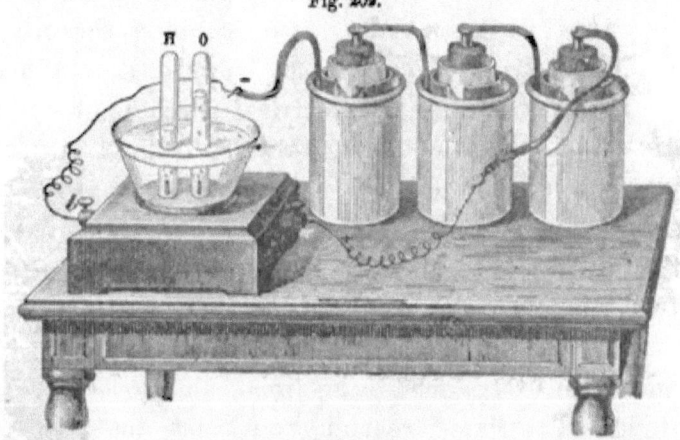

Fig. 209.

bubbles will immediately begin to rise to the surface. If the copper poles are inserted, bubbles will pass off from the negative, but none from the positive pole, since the oxygen combines with the copper wire. If the gases are collected, they will be found to be oxygen and hydrogen, in the proportion of two parts of the latter to one of the former. The theory of the change is the same as that illustrated in Fig. 205. It is a curious fact that the burning of an atom of zinc in the battery develops enough electricity to set free an atom of oxygen at the positive pole. This indicates a very intimate relation between chemical affinity and electricity—perhaps even their identity.

2. *Electrolysis* (to loosen by electricity).—This is the process of the decomposition of compound bodies by the voltaic current. A substance which, like water, can be separated in this manner, is termed an *electrolyte*.

Electro-positive and Electro-negative Substances.—In the electrolysis of compounds, their elements are found to be in different electrical conditions. Hydrogen and most of the metals go to the negative pole, and (since unlike electricities attract) are *electro-positive*. Oxygen, chlorine, sulphur, etc., go to the positive pole, and are therefore *electro-negative*. In the following list each substance is electro-negative toward those which follow it, and electro-positive toward those which precede. (Berzelius.)

Electro-negative.
1. Oxygen.
2. Sulphur.
3. Nitrogen.
4. Chlorine.
5. Iodine.
6. Phosphorus.
7. Molybdenum.
8. Tungsten.
9. Carbon.
10. Antimony.
11. Silicon.
12. Hydrogen.
13. Gold.
14. Platinum.
15. Mercury.
16. Silver.
17. Copper.
18. Bismuth.
19. Tin.
20. Lead.
21. Cobalt.
22. Nickel.
23. Iron.
24. Zinc.
25. Manganese.
26. Aluminum.
27. Magnesium.
28. Calcium.
29. Barium.
30. Lithium.
31. Sodium.
32. Potassium.

Electro-positive.

3. *Electro-typing* is the process of depositing metals from their solution by means of electricity. It is much used in copying medals, woodcuts, type, etc. An impression of the object is taken with gutta-percha or wax. The surface to be copied is brushed over with black-lead to render it a conductor. The mould is then suspended in a solution of sulphate of copper, from the negative pole of the battery; a

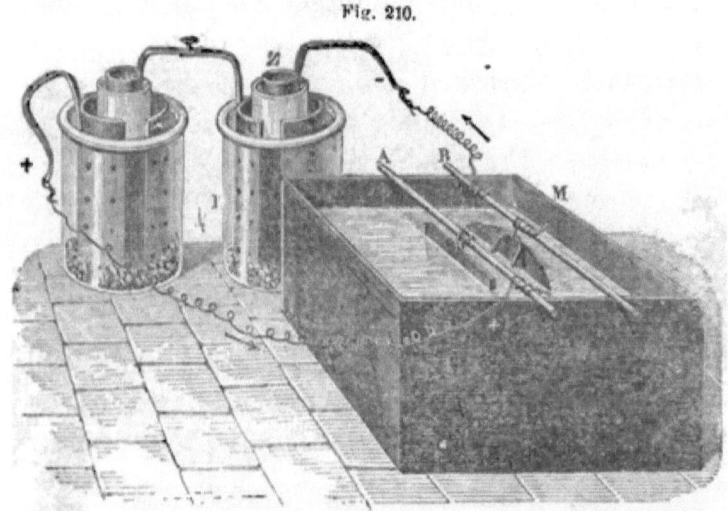

Fig. 210.

plate of copper is hung opposite on the positive pole. The electric current decomposes the sulphate of copper; the metal goes to the negative pole and is deposited upon the mould, while the acid, passing to the positive pole, dissolves the copper, and thus preserves the strength of the solution.

Duplicates of an engraved copper-plate are prepared in the following manner. The back of the plate is rendered non-conducting by a coating of varnish. The plate is then suspended in the solution. When the deposit of copper has reached the required thickness, it is stripped off without difficulty. This, of course, represents the engraved plate in relief. If a *fac-simile* is desired, a deposit is made in the same way upon the copy. Daguerreotype plates have been thus transferred without injury to the original. Leaves, insects, fruits, and even flowers, have been coated with copper by this wonderful process.

While the plate is hanging in the solution there is no noise heard or bubbling seen. The most delicate sense fails to detect any movement. Yet the mysterious electric force is continually drawing particles of *ruddy, solid* copper out of the *blue liquid*, and, noiselessly as the fall of snowflakes, dropping them on the mould; producing a metal purer than any chemist can manufacture, spreading it with a uniformity no artist can attain, and copying every line with a fidelity that knows no mistake.

4. *Electro-plating* is the process of coating with silver or gold by electricity. The metal is deposited most readily on German silver, brass, copper, or nickel silver. The last is a mixture of copper, zinc, and nickel, and is used for the best plated-ware. The vessels to be plated are thoroughly cleansed, and then hung in a solution of silver from the negative pole, while a plate of silver is suspended on the positive pole. In five minutes a mere "blush" of the metal will be depos-

ited, which perfectly conceals the baser metal and is susceptible of a high polish. It is said that an ounce of silver can in this way be spread over two acres of surface. A well-plated spoon receives about as much silver as there is in a ten-cent piece. The only method of deciding accurately the amount deposited is by weighing the article before and after being plated. A vessel is gold-lined by filling it with a solution of gold, suspending in it a slip of gold from the positive pole of the battery, and then attaching the negative pole to the vessel. The current passing through the liquid causes it to bubble like soda-water, and in a few moments deposits a thin film of gold over the entire surface.

A simple Illustration in Plating.—Place in a large test-tube a silver coin with a little aqua-fortis. If the fumes of the decomposed acid do not soon rise, warm the liquid. When the silver is dissolved fill the tube nearly full of soft water. Next drop muriatic acid into the liquid until the white precipitate (chloride of silver) ceases to fall. When the chloride has settled, pour off the colored water which floats on top. Fill the tube again with soft water; shake it thoroughly; let it settle, and then pour off as before. Continue this process until the liquid loses all color. Finally, fill with water and heat moderately, adding cyanide of potassium in small bits as it dissolves, until the chloride is nearly taken up. The liquid is then ready for electro-plating. Thoroughly cleanse a brass key, hang it from the negative pole of a small battery and suspend a silver coin from the positive pole. Place these in the silver solution, very near and facing each other. When well whitened by the deposit of silver, remove the key and polish it with chalk. In the arts the polishing is performed by rubbing with "burnishers." These are made of polished steel, and fit the surfaces of the various articles upon which they are to be used.

III. PHYSIOLOGICAL EFFECTS.—With a single cell no effect is experienced when the two poles are held in the hands. With a large battery a sudden twinge is felt, and the shock becomes painful and even dangerous, especially if the palms are moistened with salt-water, which increases the conduction. Rabbits which had been suffocated for half an hour, have been restored to life by an application of the galvanic current.

ELECTRO-MAGNETISM.

EFFECT OF A VOLTAIC CURRENT ON A MAGNETIC NEEDLE.—If a current of electricity be passed over a magnetic needle, the needle will turn and tend to place itself at right angles to the wire. If the wire be brought over and beneath the needle, it doubles the effect, and the play of the needle becomes a very delicate test of the presence and direction of the electric force.

ELECTRO-MAGNETISM. 303

Fig. 211.

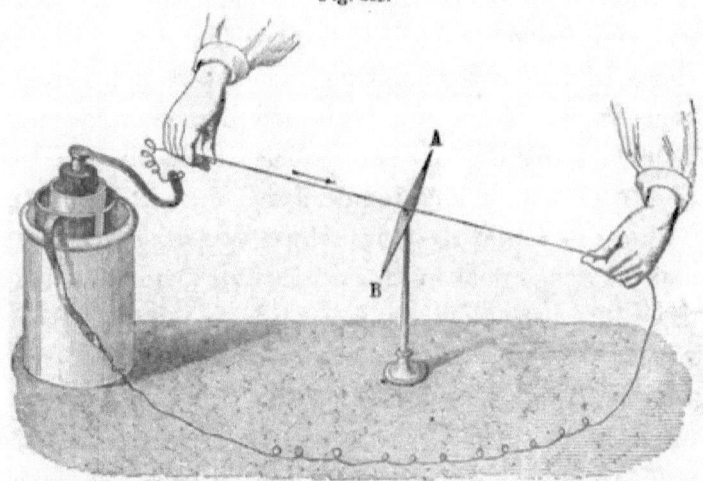

The Galvanometer is an instrument for measuring the force and direction of an electric current. B

Fig. 212.

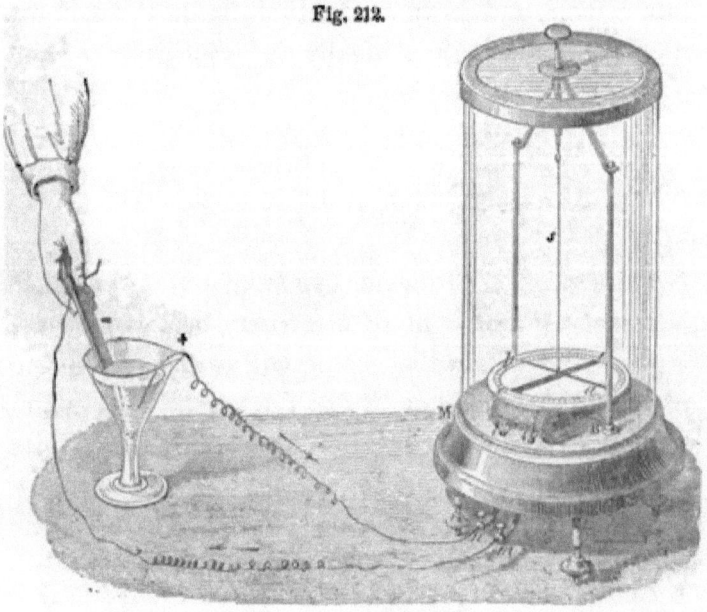

is a coil of wire, wound with thread to insulate it and compel the electricity to pass through the whole length; the current is represented as entering at *n* and leaving at *m*. The silk cord, *s*, supports an *astatic needle*. This consists of two magnetic needles, one over the graduated circle and the other within the coil, with the north pole of the one opposite the south pole of the other, so as to neutralize the attraction of the earth, and permit the combined needle to obey the will of the current alone. This affords a means of testing the faintest flow of electricity.

ELECTRO-MAGNETS.—The voltaic current produces magnetism. If a current be passed through the wire shown in Fig. 201, the steel bar will be rendered

Fig. 213.

magnetic. This shows the identity of the electricity from the voltaic battery with that from the Leyden jar. If the wire be wound around a bar of soft iron, as in Fig. 213, the iron will instantly become a magnet which will grasp the armature with great force, but will as quickly lose its properties when the current is broken. Electro-magnets have been made that would lift 3,500 times their own weight. If the current be passed through a coil of insulated wire (a *helix*), as in Fig. 214, a rod

ELECTRO-MAGNETISM.

of iron, when held below it, will be drawn up into it forcibly, as if pulled by a powerful spring; thus realizing in science the fabulous story of Mahomet's coffin, which is said to have been suspended in mid-air. Here we see that not only does the soft iron within become magnetic, but also the coil itself. Bar-magnets are now made by inserting them in a large coil through which a powerful current is passing.

Fig. 214.

Motion produced by Electricity.—If we reverse the direction of the current, it changes the poles of the magnet. Advantage is taken of this principle in order to produce continuous motion. Fig. 215 represents Page's rotating machine.

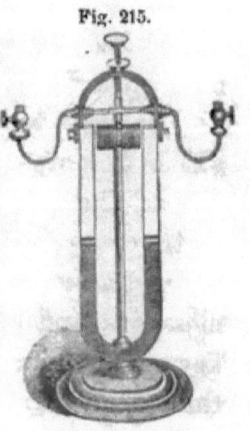
Fig. 215.

It consists of an upright horseshoe magnet, between the poles of which is a small electro-magnet. Above this are two springs, which are so placed that, as the central rod revolves with the electro-magnet, the current passes through these springs, alternately, to the wire coiled about the iron of the electro-magnet. The poles of the electro-magnet are thus changed twice with each revolution. The poles of the upright magnet attract the opposite poles of the electro-magnet, but as soon as they face each other the current is reversed, and they at once repel each other: the other poles are now attracted, but as they come together are repelled as before. A rapid motion is thus secured. The revolutions may rise as high as 2,500, making 5,000 reversals of the current in a minute.

Electro-magnetic engines are constructed either on the principle that the magnet retains its power only while the current is passing, or that the poles are changed by reversing the current. They have been made of 8 or 10 horse-power, yet have never become of great practical value, because of the expense of the battery required to produce the electricity. The zinc which burns in the cell of the electric-engine is far more expensive than the coal which burns in the furnace of the steam-engine.

The Electro-magnetic Telegraph depends on the same principle as the electro-magnet. A single wire is used to connect the two stations between which despatches are to be sent. The extremities of the wire extend into the ground, and the earth completes the circuit. Each station has a *key* and a *receiver;* the former is used for sending messages, and the latter for receiving them. The key is shown in Fig. 216. The wire, P, leads from the battery; L is the line-wire, and A connects with the receiver; a brass lever, $h\ k$, turns on an axis.

A spring, r, elevates the lever, and keeps the pin, a, pressed down upon a little button just below, to which the wire, A, is attached. The key is now in a condition to receive a message. The current from L passes through the lever k down the pin a, along the wire A, to the recording instrument, and thence to the earth, making the circuit complete. To *send* a message, the button B is pressed down by the finger of the operator, so as to strike the button below it; the circuit is established there and broken beneath a. The current from the battery at the station now passes from P through h to L. The operator, by elevating or depressing B, can thus break or complete the circuit at his option. At the station where the despatch is received, the current passes, as we have seen, directly into the receiver. This contains an electro-magnet, E. When the circuit is complete, the current, flashing through the coils of wire at E, attracts the armature, m. This elevates n, the other end of the lever, m n, and forces the sharp point, x, firmly against the soft paper, a. As soon as the circuit is broken, E ceases to be a magnet, and the spring, R, lifts the armature, drawing the point from the paper. Clock-work attached to the rollers at z moves the paper along uniformly beneath the point

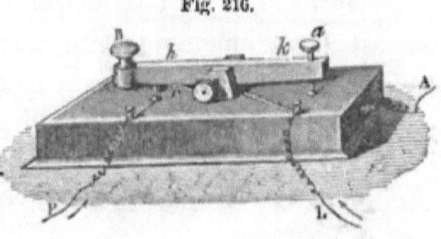

Fig. 216.

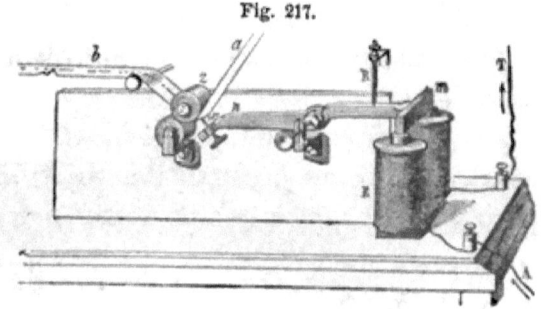

Fig. 217.

x. When the circuit is completed and broken instantly, there is a sharp dot made on the paper. This is called *e*; two dots, *i*; three dots, *s*; four dots, *h*. If the current is closed for a longer time, the mark becomes a dash; this is *t*; two dashes, *m*; a dot and a dash, *a*.

Table of Morse's Signs.

a . —	j — . — .	s . . .
b — . . .	k — . —	t —
c . . .	l —	u . . —
d — . .	m — —	v . . . —
e .	n — .	w . — —
f . — .	o . .	x . — . .
g — — .	p	y . . . .
h	q . . — .	z . . . .
i . .	r . . . ?	& . . . .

A skilful operator becomes so used to the sound that the clicking of the armature is perfectly intelligible. He uses, therefore, simply a "*sounder*," *i. e.*, a receiver without the paper and clock-work attachment. We thus see that the principle of the telegraph consists *in closing and breaking the circuit at one station, and in making and unmaking an electro-magnet at the other.*

The Relay.—When the stations are more than fifty miles apart, the current becomes too weak to work the receiver. The *relay* uses the force of a local

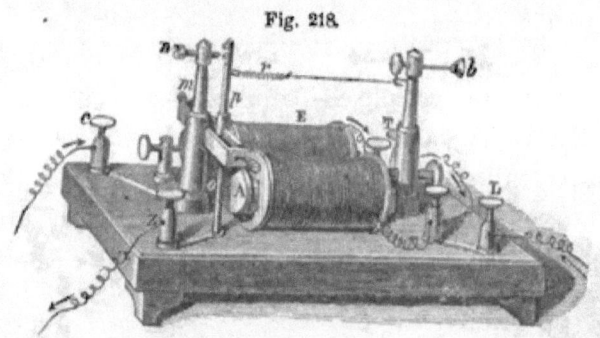

Fig. 218.

battery for this purpose. L is the line-wire; T the ground-wire; c is connected with the positive pole; Z with the receiver, and thence with the negative pole of the battery. The current passes in at L, traverses the fine wire of the electro-magnet, E, and thence passes out at T to the ground. The armature A, playing to and fro as the current from the distant station darts through or is cut off, moves the lever p, which works on an axis at its centre, and is drawn back by the spring r. As A is attracted, p strikes against the screw n; the current from C leaps up m to n, down p and through Z to the electro-magnet of the receiver and attracts its armature. The operator who sends the message simply completes and breaks the circuit with the *key*, the *ar-*

mature of the *relay*, at the station where the message is received, vibrates in unison with these movements, the *receiver* or *sounder* repeats them with greater force, and the second operator interprets their meaning.

MAGNETO-ELECTRICITY is that which is developed by means of magnetism. A common form of a machine for this purpose is shown in Fig. 219. Coils

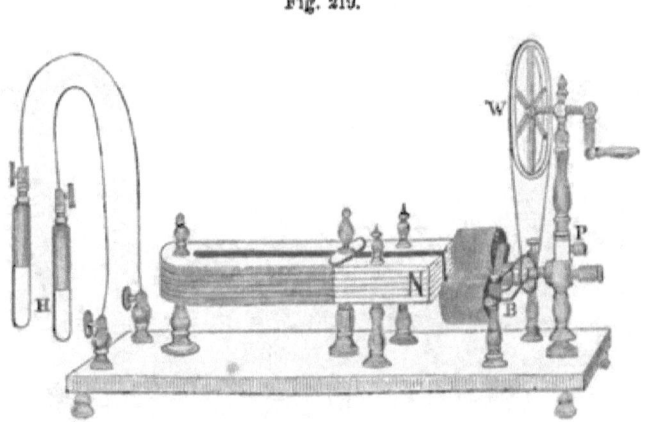

Fig. 219.

of wire are carefully insulated and wound around a small bar of soft iron, B, bent at right angles. This acts as the armature of a powerful horse-shoe magnet, before the poles of which it is made to revolve. The soft iron becomes magnetic, and then induces electric currents in the coils. The poles are changed twice, and thus two opposite currents are induced in each revolution. By means of a break-piece the circuit is rapidly broken and closed. Severe shocks are thus produced, when the poles are grasped by the hands.*

In *Wilde's machine,* the induced current from the coils is carried around a large electro-magnet, which is thereby excited to a high degree. The armature revolving before this furnishes the current which is used. A machine lately exhibited was driven by a steam-engine of 7-horse power. The poles were wire-rope, a quarter of an inch in diameter and 140 feet long. It produced an electric light dazzling as the noonday sun, throwing the flame of the street-lamps into shade at a quarter-mile distance. Its heat was sufficient to fuse a rod of iron a quarter of an inch in diameter and fourteen inches long, and could be felt fifty yards away. When one pole was inserted in the canal, and the other in a pool two hundred feet distant, the water was decomposed, oxygen gas bubbling up at one electrode, and hydrogen at the other.

INDUCED CURRENTS.—Let two coils of wire be made to fit into each other, and carefully separated by insulators. If a current of electricity be passed

---

* A Yankee once threw the industrial world of Europe into a wonderful excitement by announcing a new theory of perpetual motion based on the magneto-electric machine. He proposed to decompose water by the current of electricity, then burn the hydrogen and oxygen thus obtained. In this way he would drive a small steam-engine, which, in turn, would keep the magneto-electric machine in motion. This would certainly be a splendid discovery. It would be a steam-engine which would prepare its own fuel, and, in addition, dispense light and heat to all around. (Helmholtz.)

through the inner coil, it will induce a powerful secondary current, flowing in the opposite direction, in the outer coil. This soon ceases: on breaking the circuit, however, it will start again, but in the same direction as the primary current. The apparatus shown in Fig. 220 consists essentially of the

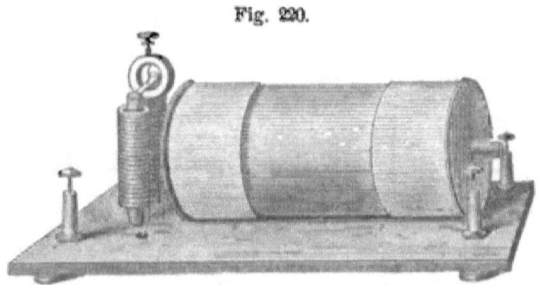

Fig. 220.

two coils just described. The primary current from a single cell is rapidly interrupted by means of a small electro-magnet. When this is magnetized, it attracts the armature, and thus the circuit is broken; the armature immediately springs back, and again completes the circuit. A bunch of iron wires may be inserted as a core in the inner coil. When the current passes, these become magnetized, and by induction largely strengthen the secondary current. This form is much used for medical purposes. *Ruhmkorff's coil* is constructed on the same principle. The largest coils often contain thirty to fifty miles of covered wire. Ritchie, of Boston, has devised many ingenious improvements which render the current extremely intense. His 15-inch coils will throw a quick succession of sparks, each fifteen inches long, charge and discharge a Leyden jar, with a crack like that of a pistol, as rapidly as one can count, and perform the vacuum experiments in frictional electricity with a splendor and brilliancy no plate-machine can rival.

THERMAL ELECTRICITY.

As electricity can be changed into heat, in turn heat can be converted into electricity. A Thermo-electric pile consists of alternate bars of antimony and bismuth soldered together, as shown in Fig. 222. When mounted for use, the couples are insulated from each other and enclosed in a copper frame P. If both faces of the pile are equally heated, there is no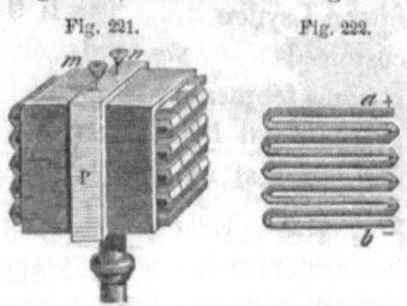
current. The least variation of temperature, however, between the two is indicated by the flow of electricity. Wires from a, the positive pole, and b, the negative, connect the pile with the galvanometer (Fig. 212). This constitutes one of the most delicate tests of the presence of heat. A tiny insect held against the face of the pile will move the needle. Strange, that minute quantities of heat become sensible only when they are converted into electricity, then into magnetism, and lastly into motion!

ANIMAL ELECTRICITY.

ELECTRIC FISH have the property of giving, when touched, a shock like that from a Leyden jar. The torpedo and electrical eel are the most noted. The former is a native of the Mediterranean, and its shock was anciently much prized as a cure for various diseases. The latter is abundant in certain South American waters. A specimen of this fish, forty inches in length, was estimated by Faraday to emit a spark equal to the discharge of a battery of fifteen Leyden jars. The Indians are said to be accustomed to drive herds of wild horses into the streams frequented by the fish. The horses are soon overpowered by the terrible shocks they receive, and so fall an easy prey to their pursuers.

CONCLUSION.

"Science is a psalm and a prayer."—PARKER.

NOWHERE in nature do we find chance. Every event is governed by fixed laws. If we would accomplish any result or perform any experiment, we must come into exact harmony with the universal system. If we deviate from the line of law by a hair's breadth, we fail. These laws have been in operation since the creation, and all the discoveries of science prove them to extend to the most distant star in space. A child of to-day amuses itself with casting a stone into the brook and watching the widening curves: little antediluvian children could have done the same. A law of nature has no force of itself; it is but *the manner in which force acts.* We cannot create force. We can only take it as a gift from God. We find it everywhere in Nature. Matter is not dumb, but full of inherent energy. A tiny drop of dew sparkling on a spire of grass is instinct with power: Gravity draws it to the earth; Chemical Affinity binds together the atoms of hydrogen and oxygen; Cohesion holds the molecules of water, and gathers the drop into a globe; Heat keeps it in the liquid form; Adhesion causes it to cling to the leaf. If the water be decomposed, Electricity would be set free; and from this, Heat, Light, Magnetism, and Motion could be produced. Thus the commonest object becomes full of fascination to the scientific mind, since in it reside the mysterious forces of Nature.

These various forces can be classified either as *attractive* or *repellant*. Under their influence the atoms or molecules resemble little magnets with positive and negative poles. They therefore approach or recede from each other, and so tend to arrange themselves according to some definite plan. "The atoms march in time, moving to the music of law." A crystal is but a specimen of "molecular architecture" built up by the forces with which matter is endowed.

No force can be destroyed. A hammer falls by the force of gravity and comes to rest, but its motion as a mass is converted into a motion of atoms, and reveals itself to the sense of touch as heat. Thus force changes its form continually, but the eye of philosophy detects it and enables us to drive it from its various hiding-places still undiminished. It assumes Protean guises, but is doubtless essentially a unit everywhere. It may disappear from the earth; still—

> "Somewhere yet that atom's force
> Moves the light poised universe."

This conversion of force is termed the "CORRELATION OF THE PHYSICAL FORCES." It is the grandest law Nature offers for the contemplation of the human mind. What is the nature of force we cannot tell. We think it to be a mode of motion. Beyond this, all is mystery.

The forces of Nature are strangely linked with our lives. Everywhere a Divine Hand is developing ideas tenderly and wondrously related to human needs. To the thoughtful mind all phenomena have a hidden meaning.

> "To matter or to force
> The all is not confined;
> Beside the law of things
> Is set the law of mind;
> One speaks in rock and star,
> And one within the **brain**,
> In unison at times,
> And then apart again.
> And both in one have brought us hither
> That we may know our whence and whither

"The sequences of law
  We learn through mind alone;
  We see but outward forms,
  The soul the one thing known;—
  If she speak truth at all,
  The voices must be true
  That give these visible things,
  These laws, their honor due,
But tell of One who brought us hither
And holds the keys of whence and whither.

"He in His science plans
  What no known laws foretell;
  The wandering fires and fixed
  Alike are miracle:
  The common death of all,
  The life renewed above,
  Are both within the scheme
  Of that all-circling love.
The seeming chance that cast us hither
Accomplishes His whence and whither."

NATIONAL SCHOOL APPARATUS.

PHILOSOPHICAL APPARATUS.
DESIGNED TO ILLUSTRATE STEELE'S 14 WEEKS IN PHILOSOPHY.

Set No. 1,—Price $125.00.

The difficulty and inconvenience of procuring the articles necessary to practically illustrate these text-books, have induced the publishers of *Peck's Ganot's, and Steele's Philosophies*, to prepare sets of Apparatus complete enough to meet the wants of all. Set No. 2 is adequate to the performance of all the principal experiments in either of the text-books mentioned, or in almost any published.

MECHANICS.
Centrifugal Hoops	$7 00
Rocking Horse	2 00
Collision Balls	12 00

ELECTRICITY.
5-in. Cylinder Machine	36 00
Leyden Jar. Qt.	3 50
Discharger	4 00
Spiral Tube	6 00
Set Bells	4 00
Plates for Images	3 00

PNEUMATICS.
Air Pump and Receiver	36 00
Hemispheres	14 00
Fountain in Vacuo	12 00
Globe to Weigh Air	3 50
Hand & Bladder Glass	2 50

HYDROSTATICS.
Equilibrium	5 00
Hydrometer and Jar	2 50
Water Hammer	1 60
Bottle Imps	1 50
Syphon	0 50
	$156 60

OPTICS.
Compound Microscope	$20 00
Concave & Convex Mirrors	4 00
Set Lenses	5 00
Magnifier	3 00
Prism	2 50

MAGNETISM AND GALVANISM.
Pot Battery	6 00
Electric Magnet	4 00
Magnet	2 00
Bar Magnet	3 00
Magnetic Needle	3 00
Dip Needle	5 00

CHEMISTRY.
Retort Stand	3 50
Spirit Lamp	2 25
6 Bulb Tubes	3 50
Barometer Tube & Mercury	4 50
6 Rupert's Drops	1 50
Funnel and Fillers	1 15
Glass Tube Assorted	1 50
Porous Cup	0 80
Glass Tubes for Sound	1 60
Compound Bar	3 00
Flask	0 60
	$81 40

Set No. 2,—Price $500.00.

Embracing all the articles in Set No. 1, with many additional fine instruments, and adequate to the performance of all the experiments in text-books generally.

These Sets are securely packed in wooden boxes, and may be safely transported to any distance. Sent by express on receipt of price, or C. O. D.

A. S. BARNES & CO.,
111 & 113 William Street, New York.

P. O. Box 1672.

NOTES

ON APPARATUS AND EXPERIMENTS.

PAGE
30. An Ivory ball from apparatus Fig. 33 can be used for this experiment.
40. Half-dozen Rupert's Drops. A glass funnel, pack of filters, and 1 lb. animal charcoal.
41. 1 lb. soft French glass tubing, assorted sizes. A 4 oz. alcohol lamp is also necessary.
44. Instead of the blue litmus, a solution of cabbage is good in this experiment. It is made by steeping purple cabbage-leaves in water, until the colored juice is extracted. The funnel can be made of tin, by any tinsmith.
46. A Grove's cup, fitted with cork and tube. This may be supported with a wire tripod or any convenient device. The hydrogen preparation is described in Chemistry.
51. Long tube for Guinea and Feather experiment. It may also be used to perform the experiment on page 282.
54. Fig. 14 can be easily made. The rocking-horse, Fig. 16 illustrates the principle more forcibly.
58. Set of Pendulums, Fig. 19.
59. This apparatus is not as essential as the last named, though very useful.
60. The apparatus, Fig. 21, can be made by any carpenter. The pendulums are turned from hard wood, and hung on wire-hooks.
71. Apparatus to explain 2d law of motion.
77. Centrifugal-force apparatus, Fig. 32.

79. Action and Reaction apparatus, Fig. 33.
85. Model of the Mechanical Powers.
104. Half-dozen tubes with bulbs, as in Fig. 67.
106. Model of Hydrostatic Press.
108. This series of tubes can be easily made by any teacher having the glass tubing and a spirit-lamp.
109. Apparatus shown in Fig. 72.
110. Apparatus shown in Fig. 73 or 74. The former is less liable to be injured by use.
117. Hydrostatic balance and weights. This instrument is furnished with glass and brass disks to estimate the adhesion of solids and liquids.
118. Hydrometer and jar. The jar may also be used in Figs. 8 and 83.
128. Model of Barker's Mill.
133. A table air-pump. This is the best and cheapest form of the air-pump. A barometer-gauge is a valuable addition. A condensing air-chamber, syringe, and jets, form a most valuable counterpart of the air-pump. By means of it, many instructive experiments in Hydraulics and Pneumatics may be performed.
133. A copper flask and stop-cock, Fig. 93.
134. Two black Cartesian imps.
136. Hand-glass. Magdeburg Hemispheres.
137. Upward-pressure apparatus.
138. Apparatus shown in Fig. 101.
138. Barometer-tube, open at both ends. 3 lbs. of mercury.
142. Model of forcing and lifting pumps.
146. A glass siphon, with tube for exhausting the air.
155. Sound in vacuo. A much cheaper apparatus than the one shown in the figure will answer the purpose of this experiment. The bell may be suspended by a cord and rung by a sliding rod, or by simply tilting the pump. The effect will not be as complete as that stated in the text.
172. Figs. 123 and 125 can be made by any ingenious pupil, and will afford profitable amusement. Try them.

173. A vibrating-plate and violin-bow, Fig. 125.
183. Glass tubes for singing flames. The experiment is most satisfactory when an apparatus like that shown in the figure is employed. The tube may, however, be held by the hand. The beaks of broken retorts make excellent tubes.
194. A concave and convex mirror in one frame.
200. A set of small lenses.
201. A large double convex lens, mounted.
205. Mounted prism. The lens just mentioned can be used for the recomposition of the light, but a more striking way is to use a painted disk, to be attached to the apparatus for centrifugal force, Fig. 32.
215. A compound microscope, with mounted objects.
219. Magic lantern. This is capable of almost unlimited use, if means can be procured to purchase mounted slides illustrative of principles in Astronomy, Geology, Botany, Physiology, etc.
234. A compound bar to illustrate unequal expansion of metals.
240. A Florence Flask, Fig. 171. This flask may be used also for Fig. 174.
240. Water-hammer or Pulse-glass.
261. Bar and horse-shoe magnets. 1 lb. iron-filings.
263. Small horizontal and dipping needles.
273. Electrical machine, electric whirl, and brass chain. An insulating stool may be extemporized with an ordinary stool, by setting the legs in glass tumblers.
274. Small insulated conductor.
276. Electric chime.
277. Dancing-image plates, and pith-ball image.
278. Leyden jar with movable coatings.
277. Leyden jar and discharger.
286. Spiral tube and diamond Leyden jar.
(All the experiments in Galvanism and Electro-magnetism can be performed with a large sulphate of copper battery, except the decomposition of water and the electric light. This battery is cheap, and very con-

venient to use. For the other experiments a battery of 5 to 12 of Grove's Cups will answer, though the electric arch cannot be well exhibited with less than 40 to 60 cups.

304. An electro-magnet.
305. A lifting coil.
305. Page's rotating machine.
307–9. Model of a telegraphic machine.
310. Magneto-electric machine in a box.
312. Electro-magnetic machine, Fig. 220.
313. Thermo-electric pile and galvanometer, Figs. 221 & 222.

☞ *Priced lists of the above apparatus will be furnished on application to*   A. S. BARNES & CO.,
111 and 113 William St., New York.

QUESTIONS.

THE following questions are those which the author has used in his own classes, both as a daily review and in examination. A standing question, which has followed every other question, has been: "*Can you illustrate this?*" Without, therefore, a particular request, the pupil has been accustomed to give as many practical examples as he could, whenever he has made any statement or given any definition. The figures refer to the page of the book.

INTRODUCTION.—Define matter. A body. A substance. Name and define the two kinds of properties which belong to each substance.

14. The two kinds of change. What is the principal distinction between Philosophy and Chemistry? Mention some phenomena which belong to each. Why are these branches intimately related?

15. Name the general properties of matter. Define magnitude. Size. Why is it necessary to have a standard of measure? Whence were the ancient standards derived? Give the history of the English standard.

16. Is the American yard an exact copy of the English? Have we any national standard? Give an account of the French system. By what name is this system commonly known? Is either of these systems founded on any natural standard? Why is it desirable to have such a standard?

17. Define Impenetrability. Give some apparent exceptions, and explain them. Define Divisibility.

19. Is there any limit to the divisibility of matter? Explain the Atomic Theory. What use has it?

20. How do animalculæ illustrate this subject? Under a powerful microscope how would chalk-marks appear?

21. Define Porosity. Is the word *porous* here used in its common acceptation? Define a molecule. An atom. Compare the size of an atom with that of a pore.

25. Define Inertia. Does a ball, when thrown, stop itself? Why is it difficult to start a heavy wagon? Why is it dangerous to jump from the cars when in motion?

26. Define Indestructibility. Did the earth, at its creation, contain the same quantity of matter it does now?

27. Name the specific properties of matter. Define **Ductility**. How is iron wire made?

28. Platinum wire? Gilt wire? What is said of brass wire? Define Malleability. Describe the manufacture of gold-leaf.

29. Is copper malleable? Define Tenacity. Name and define the three kinds of Elasticity. Illustrate the elasticity of compression as seen in solids.

30. In liquids. In gases. What is said about the relative compressibility of liquids and gases? Compare air with water.

31. Illustrate the elasticity of expansion as seen in solids, liquids, and gases. Define Elasticity of Torsion. What is a Torsion balance? Define Hardness. Does this property depend on density?

32. Define Density. Define Brittleness. Is a hard body necessarily brittle? Why are feathers light and lead heavy?

MOLECULAR FORCES.—Define a molecular force. What two opposing forces act between the molecules of matter? How is this shown? What is the repellant force? Name the attractive forces. Which of these belong to Philosophy?

COHESION.—Define. What are the three states of matter? Define. How can a body be changed from one state to another? Show that cohesion acts only at insensible distances. Explain the process of welding.

37. Why cannot all metals be welded? Why do drops of dew, etc., take a globular form? Why do not all bodies have this form?

38. Illustrate the tendency of matter to a crystalline structure. Has each substance its own form?

39. Why is not cast-iron crystalline? Why do the axles of cars become brittle after use? Describe the process of tempering and annealing.

40. Explain the Rupert Drop.

ADHESION.—Define. What is the theory of filtering through charcoal?

41. Of what use is soap in making bubbles? Define Capillary Attraction. Why will water rise in a glass tube, while mercury will be depressed? Is a tube necessary to show capillary attraction? What is the law of the rise in tubes?

42–3. Give practical illustrations of capillary action. Why will not old cloth shrink as well as new, when washed?

44. What is the cause of solution? Why is the process hastened by pulverizing? Tell what you can about gases dissolving in water. Why does the gas escape from soda-water as soon as drawn? Why do pressure and cold favor the solution of a gas? Describe the diffusion of liquids.

45–7. Of gases. The osmose of liquids. Of gases. Why do rose-balloons lose their buoyancy?

GRAVITATION.—How does Gravitation differ from Cohesion and Adhesion? What is the law of gravitation? Why does a stone fall to the ground? Will a plumb-line near a mountain hang perpendicularly? Why do the bubbles in a cup of tea gather on the side?

49. How is the earth kept in its place? Define Gravitation. Gravity. Weight. Give the three laws of weight.

50–2. What is a vertical or plumb-line? Give the four laws of falling bodies. Describe the "guinea and feather experiment." What does it prove?

53. Give the equations of falling bodies. How can the time of a falling body be used for determining the depth of a well? How does gravity act upon a body thrown upward? What velocity must be given to a ball to elevate it to any point? How high will it rise in a given time? When it falls, with what force will it strike the ground?

54–6. Define the Centre of Gravity. The line of direction. The three states of equilibrium. How may the centre of gravity be found? Give the general principles of the centre of gravity. Describe the leaning tower of Pisa.

57. Give some physiological applications of the centre of gravity. Why do fat people always walk so erect?

58–9. Define the Pendulum. Arc. Amplitude. What are isochronous vibrations. Give the four laws of the pendulum. Who discovered the first law? How?

60–2. What is the centre of oscillation? How is it found? Describe the pendulum of a clock. How is a clock regulated? Does it gain or lose time in winter? Describe the gridiron pendulum.

63. Name the various uses of the pendulum.

MOTION.—Define motion, absolute and relative. Rest. Velocity. Force. What are the resistances to motion? Tell what you can about friction. Why does oil diminish friction?

68. What uses has friction? What law governs the resist-

ance of air or water? What is the striking force? (p. 81, Prob. 35.) What is the tendency of gravity? Define Momentum.

69. Show that motion is not imparted instantaneously.

70-1. Give the three laws of motion and the proof of each. If a ball be fired into the air when a horizontal wind is blowing, will it rise as high as if the air were still? Define compound motion.

72. Define the "parallelogram of forces." The resultant. How can the resultant of two or more forces be found? Give practical illustrations of compound motion.

73. What is the "resolution of forces?" Show how one vessel can sail south and another north, driven by the same westerly wind.

74-5. Explain how a kite is raised. Explain the towing of a canal-boat. Define circular motion.

76. Apply the principle of circular motion to the revolution of the earth about the sun.

77. Show when the centrifugal force becomes strong enough to overcome the force of Cohesion, Adhesion, Gravity. What effect does the revolution of the earth on its axis have upon all bodies on the surface? What would be the effect if the rotation were to cease? Describe the action of the centrifugal force on a hoop rapidly revolved on its axis.

78. Give practical illustrations of action and reaction. If a bird could live, could it fly in a vacuum?

79. Define reflected motion. Give its law.

80. How is curved motion produced? Is perpetual motion practicable?

THE MECHANICAL POWERS.—Name and define the elements of machinery. Do the "powers," so called, produce force? What is the law of Mechanics? Illustrate the law.

86-7. Describe the three classes of levers. The law of equilibrium.

88. What is the advantage peculiar to each class? Describe the steelyard as a lever. What effect does it have to reverse the steelyard?

89. Describe the arm as a lever. Would a lever of the first class answer the purpose of the arm? What is a bent lever?

90. Describe the compound lever. The wheel and axle.

91. The capstan. Give the law of equilibrium. What is the advantage of the wheel and axle?

92. Describe a system of wheel-work. At which arm of the lever is the P. applied?

93-4. Describe the various uses of the inclined plane. Its law of equilibrium. What velocity does a body acquire in rolling down an inclined plane? Give illustrations.

95. Describe the screw. Its uses. Its law of equilibrium How may its power be increased? What limit is there?

96. Describe the wedge. Its uses. Its law of equilibrium. How does it differ from that of the other powers?

97. Describe the pulley. The use of fixed pulleys. Is there any gain of P. in a fixed pulley?

98. The use of a movable pulley. Describe a movable pulley as a lever.

99. Give the general law of equilibrium in a combination of pulleys. What part of the force is lost by friction?

HYDROSTATICS.—Define. What liquid is taken as the type? What is the first law of liquids? Explain. Illustrate the transmission of pressure by water.

105. Show how water is used as a mechanical power.

106. Describe the hydrostatic press. Give its law of equilibrium.

107. What are the uses of this press? What pressure is sustained by the lower part of a vessel of water, when acted on by gravity alone? How does this pressure act?

108. Give the four laws which depend on this principle, and illustrate them. What is the weight of a cubic foot of sea-water? Fresh water? What is the pressure at two feet?

109. Give illustrations of the pressure at great depths.

110. Describe the hydrostatic bellows. Its law of equilibrium.

111. What is the hydrostatic paradox? Give illustrations. Give the principle of fountains. How high will the water rise?

112-3. How do modern engineers carry water across a river? Did the ancients understand this principle? Give the theory of the Artesian well.

114. Give the rule for finding the pressure on the bottom of a vessel. On the side.

115. Define the water-level. Is the surface of water horizontal? If it were, what part of an approaching ship would we see first? Describe the spirit-level. Define specific gravity. What is the standard for solids and liquids? For gases?

116-7. Explain the buoyant force of liquids. What is

Archimedes's law? Describe the "cylinder and bucket experiment." What does it prove?

118. Give the method of finding the specific gravity of a solid. A liquid. Suppose the solid is lighter than water and will not sink, what can you do? *Ans.* Tie a heavy solid to it, and then make allowance for this in calculating the specific gravity. Explain the hydrometer.

119. How can you find the weight of a given bulk of any substance? The bulk of any given weight? The exact volume of a body?

120-1. Illustrate the action of dense liquids on floating bodies. Why will an iron ship float on water? Where is the centre of gravity in a floating body? How do fish sink at pleasure?

HYDRAULICS —Define. To what is the velocity of a jet equal? How is the velocity found? Give the rule for finding the quantity of water which can be discharged from a jet in a given time.

124. What is the effect of tubes? Tell something of the flow of water in rivers.

125-7. Name and describe the different kinds of waterwheels. Which is the most valuable form? What is the principle of the Turbine? Describe Barker's Mill.

128-9. How are waves produced? Explain the real motion of the wave. How does the motion of the whole wave differ from that of each particle? How is the character of waves modified near the shore?

130. What is the extreme height of "mountain waves?" Define like phases. Unlike phases. A wave-length. What is the effect if two waves with like phases coincide? With unlike phases? What is this termed?

PNEUMATICS.—Define. What principles are common to liquids and gases? What gas is taken as the type? Describe the air-pump. Can a perfect vacuum be obtained in thi, way? Prove that the air has weight.

134-5. Show its elasticity and compressibility. Describe the bottle-imps. What principles do they illustrate? Show the expansibility of the air.

136. Describe the experiments with the hand-glass. The Magdeburg hemispheres. What do they prove?

137. Show the upward pressure of the air. The buoyant force of the air. Would a pound of feathers and a pound of

lead balance, if placed in a vacuum? On what principle does a balloon rise?

138–140. What is the amount of the pressure of the air? Describe the experiment illustrating this. Where do these figures apply? Describe how the pressure of the air constantly varies.

141. Give Mariotte's law. Describe the barometer. Its uses. Are the terms "fair," "foul," etc., often placed on the scale, to be relied upon?

142. Why is mercury used for filling the barometer? Describe Otto Guericke's barometer.

143–4. Describe the action of the lifting-pump. The force-pump. The fire-engine.

145–6. The siphon. Explain its theory.

147. The pneumatic inkstand. What was the view of the ancients concerning the pressure of the air? Tell the story of Galileo. What opposing forces act on the air?

148. How high does the air extend? How does its density vary?

ACOUSTICS.—Define. Name and define the two senses of this word. May not "light," "heat," etc., be used in the same way? Illustrate the formation of sound by vibrations.

152. Show how the sound of a tuning-fork is conveyed through the air. Report of a gun. The sound of a bell. The human voice. Define a sound-wave. A wave-length. In which direction do the molecules of air vibrate? In what form do the waves spread? Can a sound be made in a vacuum?

155. Can a sound come to the earth from the stars? How do sounds change as we pass above or below the sea-level? Upon what does the velocity of sound depend? Why is this?

156. At what rate does sound travel in the air? In water? In iron? What effect does temperature have on the velocity of sound? Describe Biot's experiment in the water-pipes of Paris. Do all sounds travel at the same rate?

157–8. How does the velocity of sound enable us to determine distance? Upon what does the intensity of sound depend? At what rate does it diminish? Why? State wherein the laws of sound are similar to those of other phenomena. What does this uniformity prove? Explain the speaking-tube.

159. The ear-trumpet. The speaking-trumpet. What is the refraction of sound?

160. Define reflection of sound. What is the law? Give

curious instances of reflection. What is the shape of a whispering-gallery?

161. Illustrate the decrease of sound by repeated reflection. Why are sounds more distinct at night than by day? What is a resonance?

162. Is it desirable to have a door or a window behind a speaker? What causes the "ringing" of a sea-shell? When is an echo heard? When is the echo repeated?

163. What is the difference between noise and music? Upon what does pitch depend?

164–6. Describe the siren. How is it used to determine the number of vibrations in any sound? How is the octave of any note produced? How can we ascertain the length of the wave in sound?

167. What length of wave produces the low tones in music? The high tones? Give the illustration of the locomotive whistle. When are two tones in unison? How can we find the length of the wave in *any* musical sound?

168. What is the length of the wave in a man's voice in common conversation? How can two sounds produce silence? What is this effect termed?

169. Illustrate interference by means of a tuning-fork. Describe the vibration of a cord.

170–1. Describe the sonometer. What is the object of the wooden box? Give the three laws of the vibration of cords. What is a node?

172. Describe the experiments illustrating the formation of nodes.

173. What are acoustic figures? Nodal lines?

175. What is the fundamental tone of a cord? A harmonic? What causes the difference in the sound of various instruments? Does a bell vibrate in nodes? The violin-case? A piano sounding-board?

176. Give the fractions representing the relative rates of vibration of the different notes of the scale. How is the sound produced in wind-instruments?

177. How is the sound-wave started in an organ-pipe? In a flute? What determines the pitch?

178. What are sympathetic vibrations? Describe the ear.

179. What is the object of the Eustachian tube? Is there any opening between the external and internal ear? What effect does it have on the hearing to increase or diminish the

pressure of the air? How does a concussion sometimes cause temporary deafness? How can this be remedied?

180. What are the limits of hearing? Does the range vary in different persons? What sounds are generally most acutely heard?

181. Are there probably sounds in nature we never hear? Has nature a tendency to music? What causes the "whispering of the pines?"

182. What is the key of nature? What are sensitive flames? How can a flame be made to sing? What causes the song?

OPTICS.—Define. A luminous body. A non-luminous body. A medium. A transparent body. A translucent body. An opaque body. A ray of light. Show that neither air nor water is perfectly transparent. Why is the sun's light fainter at sunset than at mid-day?

188. Define the visual angle. Show how distance and size are intimately related. Give the laws of light. Do they resemble those of sound?

189. What is the velocity of light? How is this proved? Explain the undulatory theory of light.

190. How does light-motion differ from sound-motion? What is diffused light? Why are some objects brilliant and others dull? Why can we see a rough surface at any angle, and an image in the mirror at only a particular one? Would a perfectly smooth mirror be visible? How does reflection vary? Define mirrors. Name and define the three kinds. What is the action of each on rays of light? What is the general principle of mirrors?

192-3. Why is an image in a plane mirror symmetrical? Why is it reversed right and left? Why is it as far behind the mirror as the object is before it? If you sit where you cannot see another person's image, why cannot that person see yours? Why can we often see in a mirror several images of an object? Why can we see these best if we look into the mirror very obliquely? Why is an image seen in water inverted?

194. When the moon is near the meridian why can we see the image in the water at only one spot? When do we see a tremulous line of light? Define the focus. Centre of curvature.

195. Describe the image seen in a concave mirror. Why is it inverted when we stand between the centre of curvature and the principal focus? Why is it larger than life when we stand within the principal focus, and smaller than life when we stand without the centre of curvature?

196. What are conjugate foci? Describe the image seen in a convex mirror. Why is it smaller than life? Why can it not be inverted like one seen in a concave mirror? *Ans.* Because the rays do not cross each other.

197. Define total reflection. Define Refraction. Does the partial reflection of light as it passes from one medium to another of different density have a parallel in sound? Why is powdered ice opaque while a block of ice is transparent? Give illustrations of refraction.

198. Why does an object in water appear to be above its true place? What is the general principle of refraction?

199. Give the laws of refraction. Describe the path of a ray through a window-glass. Is the direction of objects changed? Describe the path through a prism.

200. Name and describe the different kinds of lenses. What is the effect of a double convex lens on rays of light?

201. What is this kind of lens often called? Describe the image. Why is it inverted after we pass the principal focus? Why is it decreased in size?

202. What is the effect of a double concave lens on rays of light? Describe the image. Why can it not be inverted like one through a double convex lens? Describe the images seen in the large vases in the windows of drug-stores. What is a mirage?

203. Give its cause.

204. How is the solar spectrum formed? Name the seven primary colors. Show that these seven will form white light. What other opinions are held?

205. Why are the rays separated? What is meant by the dispersive power of a prism? What substance possesses this property in the highest degree?

206. What three classes of rays compose the spectrum? Do artificial lights differ in their proportion of these rays? What color, also, predominates? *Ans.* Yellow. Why does the window of a photographer's dark room sometimes contain yellow glass? Define complementary colors. How can they be seen? What is the effect of complementary colors when brought in contrast? (In Fig. 153, opposite colors are complementary.) Ought a red flower to be placed in a bouquet by an orange one? A pink or blue with a violet one? Why do colors seen by artificial light appear differently than by daylight—as yellow seems white, blue turns to green, etc.?

207-8. Describe Newton's rings. How are these explained

according to the wave-theory? What can you say about the length of the waves?

209. State the analogy between color and pitch in music. Why is grass green? When is a body white? Black? What causes the play of color in mother-of-pearl? In soap-bubbles? In the scum on stagnant water? In thin layers of mica or quartz? What is a tint?

210. Define diffraction. What is double refraction? What are the two rays termed? What is polarized light?

211. How does a dot appear through Iceland spar? What other methods of polarizing light? Give some illustrations and practical uses of polarized light.

212-3. How is the rainbow formed? Why must it rain and the sun shine at the same time, to produce the bow? Why is the bow in the sky opposite the sun? How many refractions and reflections form the primary bow? The secondary? How many colors can one receive from a single drop? Why is the bow circular?

214. How are halos formed? What is the cause of the "sun's drawing water?" Explain spherical aberration.

215. Chromatic aberration. Its remedy. What is the meaning of the word microscope? Describe the simple microscope. The compound microscope. How is the power of a microscope indicated? Do we see the object directly in a microscope? Why is the object-lens made so small and so convex?

216-8. What is the meaning of the word telescope? Describe the reflecting telescope. The refracting telescope. What is the use of the object-lens? The eye-piece? Is the image inverted? Describe the opera-glass.

219. The stereoscope. The magic lantern. How are dissolving views produced?

220. Describe the Camera. The structure of the eye.

222. The formation of an image on the retina. The adjustment of the eye. The cause of near and far sightedness. The remedy. Why do old people hold a book at arm's-length?

223. Illustrate the duration of an impression. Why are we not sensible of darkness when we wink? Why can we not see the fence-posts when we are riding rapidly? Describe color-blindness. What is the range of the eye?

HEAT.—Define luminous heat. Obscure heat. A diathermanous body. Cold. Gases and vapors. Show the intimate relation between light and heat.

228-9. What is light? How do the three classes of rays in the solar spectrum differ? What effect does each of these produce? What is the theory of heat? Why can we not see with our fingers or taste with our ears? At what rate does nerve-motion travel? How long does it take a tall man to find out what is going on in his foot? What is meant by the quality of heat? Does this find any analogy in sound?

230. Name the sources of heat. Describe and illustrate each of these. Can force be destroyed? If apparently lost, what becomes of it? What is Joule's law?

232. Define latent, sensible, and specific heat.

233. Explain the paradox, "that freezing is a warming process and thawing a cooling one." Explain the action of a freezing mixture. Why does heat expand and cold contract? What do you say as to the uniformity of the expansion of solids, liquids, and gases?

234. Illustrate the expansion of solids. Is it better to buy alcohol in summer or in winter?

235. What is the thermometer? Describe it. Describe the process of filling and grading.

236. The F., C., and R. scales. Tell what you can about liquefaction. Of a solid. Of a gas. In one case sensible heat becomes latent, in the other latent heat becomes sensible—why is this?

237. Give the theory of vaporization. Distillation. Since rain comes from the ocean, why is it not salt?

238-9. Describe the theory of boiling. What is the boiling point? Do all liquids boil at the same temperature? What would be the effect, if this were the case? Upon what does the boiling-point depend? Why does salt-water boil at a higher temperature than fresh-water? Why will milk boil over so easily? Why will soup keep hot longer than boiling water? Does the air, dissolved in water, have any influence on the boiling-point? (Page 247.) Can you measure the height of a mountain by means of a tea-kettle and a thermometer? Show how cold water may be used to make warm water boil.

240. At what temperature will water boil in a vacuum? Why? Describe the water-hammer and the pulse-glass.

241. Can we heat water in the open air above the boiling-point? What becomes of the extra heat? What is the latent heat of water? Upon what principle are buildings heated by steam? Have you ever seen any steam? Define evaporation. Does snow evaporate in the winter? What can be done to

hasten evaporation? Why is a saucepan made broad? Why do we cool ourselves by fanning? Why does an application of spirits to the forehead allay fever? Why does wind hasten the drying of clothes? Describe a vacuum-pan. Why is evaporation hastened in a vacuum?

242. Why is evaporation a cooling process? How is ice manufactured in the tropics? What is the spheroidal state?

243. Name and define the three modes of communicating heat. Give illustrations showing the relative conducting power of solids, liquids, and gases. What substances are the best conductors?

244. Is water a good conductor? Air? What is the principle of ice-houses? Fire-proof safes? Why do not flannel and marble appear to be of the same temperature? Is ice always of the same temperature, or is some ice colder than others? Describe the convective currents in heating water. Where must the heat be applied? Where should ice be applied in order to cool water?

245. Describe the convective currents in heating air. Upon what principle are hot-air furnaces constructed? Ought the ventilator at the top of a room to be opened in winter? At the bottom? Is space warmed by the sunbeam? Show how the glass in a hothouse acts as a trap to catch the sunbeam. Does the heat of the sun come in through our windows? Does the heat of our stoves pass out in the same way? *Ans.* It does, but only through absorption by the glass, and not by direct radiation from the fire. Show how the vapor in the air helps to keep the earth warm. The top of a mountain is nearer the sun, why is it not warmer? (Page 249.) Why does ice form at night on the Desert of Sahara? Explain the relation between absorption and reflection. Is a dusty boot hotter to the foot than a polished one? What is the elastic force of steam at the ordinary pressure of the air? What is the difference between a high-pressure and a low-pressure engine? Which is used for a locomotive? Why?

247. Describe the governor. What is the object of a fly-wheel?

248. How does the capacity of the air for moisture vary? What is the principle on which dew, rain, etc., depend? Show that a change in density produces a change in temperature. What effect does this have on the temperature of elevated regions?

250. How is dew formed? Upon what objects will it collect

most readily? Why will it not form on windy nights? Is a heavy dew a sign of rain? *Ans.* Yes, because it shows that the moisture of the air is easily condensed. The "sweating of a pitcher?" What is frost? *Ans.* Frozen dew. Why will a slight covering protect plants from frost? *Ans.* Because it prevents radiation. Why is there no frost on cloudy nights? *Ans.* The clouds act like a blanket, to prevent radiation and keep the earth warm. What is a fog?

251. How does a fog differ from a cloud? Why are mountains "cloud-capped?" Why do clouds remain suspended in the air, contrary to gravity?

252. Describe the different kinds of clouds. Describe the formation of rain.

253-4. Snow. Winds. Land and sea breezes. Tradewinds. Oceanic currents. Tell about the Gulf Stream. Explain the influence which water has on climate. Of what practical use is the air in water?

256. Describe the exception which exists in the freezing of water. Why is this made? Describe the two processes by which pure water can be obtained. How is an excessive deposit of dew prevented?

ELECTRICITY.—Give the origin of this word. Name the different kinds of Electricity. Define Magnetism. A Magnet. A natural magnet. An artificial one. A bar-magnet.

262. A horse-shoe magnet. The poles. The magnetic curves.

264. Describe a magnetic needle. What is the law of magnetic attraction and repulsion? Define magnetic induction. Explain it. When is a body polarized? Give some illustrations of induced magnetism.

265. Does a magnet lose any force by induction? How do you explain the fact that if you break a magnet each part will have its N. and S. pole? Describe the process of making a magnet. On what principle will you explain this?

266. Describe the compass. Is the needle true to the pole? What causes it to vary? What is the line of no variation? Declination?

267. Why does the needle point N. and S.? What is a dipping-needle? Explain. How is a needle balanced?

268. Where is the N. magnetic pole? How could one know when he reaches it? Does the earth induce magnetism? Which end of an upright bar will be the S. pole? How has the lodestone become polarized?

269. Define frictional electricity. The electroscope. Dif-

ference between static and dynamic electricity. Show the existence of two kinds of electricity. Give the names applied to each.

271. State the law. What is the theory of electricity? Is it a polar force? Is it easily disturbed? Define a conductor. An insulator.

272-3. What is the best conductor? Best insulator? Is a poor conductor a good insulator? When is a body said to be insulated? Can electricity be collected from an iron rod? Describe an electrical-machine. What is the use of the chain in the negative pole?

274. Define electrical induction.

275. Faraday's theory.

276. Describe the electric chime. Explain.

277. The dancing images. The Leyden jar. What gives the color to the spark?

278. How is the jar discharged? What are the essentials of a Leyden jar? What is the object of the glass? The tin-foil?

279. Give the theory of the charging of the jar. Can an insulated jar be charged? Is the electricity on the surface or in the glass? Can the inner molecules of a solid conductor be charged? Will a rod contain any more electricity than a tube? Why is the prime conductor of an electrical-machine hollow? What is the effect of points? How can we test this?

280. Describe the electric whirl. Explain the existence of electricity in the atmosphere.

281. What is the cause of lightning? Thunder? Is there any danger when you once hear the report? Describe the different kinds of lightning. Tell how Franklin discovered the identity of lightning and frictional electricity.

282. What is the cause of the Aurora Borealis? How is this shown? Prove the intimate relation between the aurora and magnetism. What are Geissler's tubes? Gassiot's cascade?

283. Tell what you can about lightning-rods.

284. In what consists the main value of the rod? Does the lightning ever pass upward from the earth? *Ans.* It does, both quietly and by sudden discharge. Has Nature provided any lightning-rods? What is St. Elmo's fire? What is the velocity of electricity?

285-6. Illustrate its instantaneousness. Name some of the effects of frictional electricity—(1) Physical, (2) Chemical, (3) Physiological.

287. How are galvanic electricity and chemistry related?

Why is galvanic or voltaic electricity thus named? Tell the story of Galvani's discovery. What was his theory?

288. Give an account of Volta's discovery. What was his theory? How can we form a simple pile? Describe the simple galvanic circuit.

289. Define the poles. Electrodes. Closing and breaking the circuit. What is necessary to form a voltaic pair?

290. Are the terms applied to the metals the same as those to the poles? Give the chemical change. Why does the hydrogen come off from the copper?

291. Tell what you can about the current. What really passes along the wire? How is this force transmitted? Will a tube, then, *convey* as much electricity as a rod?

292. Describe Smee's battery. Grove's battery. The chemical change.

293. The advantages of Grove's battery. Describe Bunsen's battery. Daniell's battery.

294. Sulphate of copper battery. Define quantity and intensity. Upon what do they depend? Compare frictional and galvanic electricity.

295. Give the effects of galvanic electricity, (1) Physical—heat and light; (2) Chemical—decomposition of water, electrolosis, electrotyping, duplicates of copper-plates, and electro-plating; (3) Physiological.

302-3. What is the effect of a voltaic current on a magnetic needle? What is a galvanometer?

304. An astatic needle? An electro-magnet? A helix?

305. Show how a helix can be magnetized. How are bar-magnets made? How is motion produced by electricity? Describe Page's rotating-machine.

306. What is the principle of an electric engine? What difficulty is there in the way of its practical use? Describe the magnetic telegraph.

307. How is a message sent? How is one received?

308-9. What is a sounder? What is the general principle of the telegraph? Describe the relay. Name the use of each instrument.

310-11. Define magnetic electricity. Describe a magneto-electric machine. Describe Wilde's machine. Induced currents.

312. Ruhmkorff's coil.

313. Thermal electricity. A thermo-electric pile.

314. Describe the electric fish.

INDEX.

Acoustics..... 151
" figures 173
Action and Reaction... 78
Adhesion..... 40
Air..... 132
Air-pump..... 132
Alcoholmeter..... 118
Amplitude..... 58
Animalculæ..... 20
Annealing..... 30
Arm, The..... 89
Artesian Wells..... 112
Atmosphere..... 132
Atomic Theory..... 19
Attraction..... 33
" of Adhesion. 40
" Cohesion . 36
" Capillary... 41
" Gravitation . 48
Aurora..... 282

Barker's Mill..... 127
Barometer..... 141
Battery, Bunsen's..... 293
" Grove's..... 292
" Sulphate of Copper..... 293
" Thermo-electric..... 313
Bell..... 175
Boiling..... 238, 244
Britt cues..... 32

Camera..... 230
Capillarity..... 41
Capstan..... 91
Cartesian Diver..... 134
Centre of Gravity..... 54
" Oscillation ... 60
Chemical Affinity..... 35
Chromatic Aberration.. 215
Clock..... 61, 64
Clouds..... 251
Cohesion..... 36
Coils, Induction..... 312
Color..... 209
" -blindness..... 223
" Prismatic..... 204
" Complementary.. 206
Compass..... 266
Compensation Pendulum..... 62
Compressibility..... 26
Conductors..... 271
Cords..... 109
Correlation of Forces.. 310
Crystals..... 38, 315
Current, Voltaic..... 291
" of rivers..... 124
Curves, Magnetic..... 263

Declination..... 266
Density..... 32
Dew..... 250

Diamond Jar..... 286
Diathermancy..... 227
Diffraction..... 210
Diffusion of Liquids.... 44
" Gases..... 45
Distillation..... 237
Divisibility..... 17
Ductility..... 27

Ear, The..... 178
Ear-trumpet..... 159
Echoes..... 161
Elasticity..... 29
Electric Battery..... 292
" Light..... 296
" Telegraph..... 306
" Whirl..... 280
Electrical-machine..... 273
Electricity..... 259
" Frictional... 269
" Galvanic... 287
" Magnetic... 264
Electrodes..... 289
Electro-gilding..... 300
" -magnetism..... 302
" -magnets..... 304
" -negative and positive substances..... 298
" -plating..... 300
Electrolysis..... 298
Electrolyte..... 298
Electroscope..... 260
Equilibrium..... 54
Evaporation..... 241
Expansion..... 233
Eye, The..... 220

Falling Bodies..... 50
Far-sightedness..... 222
Filtering..... 23
Fire-engine..... 145
Flames, Sensitive..... 182
" Singing..... 182
Floating Bodies..... 119
Focus..... 200
Fogs..... 250
Force..... 67
" Pump..... 144
" Centrifugal..... 75
" Centripetal..... 75
" Composition of... 72
" Molecular..... 35
" Resolution of... 73
Fountains..... 112
Freezing Mixture..... 233
" of Water..... 236
Friction..... 67

Galvanometer..... 303
Gases..... 227
" Adhesion of..... 44
" Buoyancy of..... 137
" Compressibility.. 26, 30

Gases, Diffusion of..... 45
" Elasticity of..... 29
" Osmose of..... 44
" Pressure of..... 132
Gassiot's Cascade..... 283
Geissler's Tubes..... 283
Gold Leaf..... 29
Governor, The..... 247
Gravitation..... 48
Gravity..... 49
" Centre of..... 54
" Specific..... 115
Gulf Stream..... 254

Halos..... 214
Hardness..... 31
Harmonics..... 175
Heat..... 225
" affected by Rarefaction..... 248
" Absorption of... 246
" Conduction of... 243
" Convection of... 245
" Expansion by... 233
" Latent..... 232
" Luminous..... 227
" Mechanical Equiv. 231
" Quality of..... 229
" Radiation of..... 245
" Reflection of..... 246
" Refraction of..... 228
" Solar..... 230
" Specific..... 230
" Theory of..... 228
" Vaporization..... 237
Heating by Steam..... 241
Helix..... 304
Horse-power, A..... 100
Hydraulics..... 122
Hydrometer..... 118
Hydrostatics..... 103
Hydrostatic Bellows..... 110
" Paradox..... 111
" Press..... 106

Iceland Spar..... 211
Inclined Plane..... 93
Indestructibility..... 23
Induction..... 294, 311
Inertia..... 25
Insulators..... 271
Isochronous..... 58

Joule's Law..... 232

Kite..... 75

Lenses..... 200
Land and Sea Breeze... 258
Lever..... 85
Leyden Jar..... 277
Light..... 185
" Composition of... 204
" Diffraction of.... 210

340 INDEX.

Light, Interference of.. 207
" Laws of 188
" Polarized. 210
" Reflection of.... 199
" Refraction of... 197
" Theory of....... 189
" Total Reflection 197
" Velocity of. 189
" Waves of....... 208
Lightning............. 281
Liquids, Buoyancy of... 116
" Cohesion of.... 36
" Compressibility of......... 26, 29
" Diffusion of.. 44
" Elasticity of. 29
" Osmose of... 45
" Pressure of.... 114
" Specific Gravity of........ 118
" tend to spheres 37
Liquefaction.......... 236

Machinery........ . 85
Magdeburg Hemisphere 136
Magic Lantern........ 219
Magnetic Curves...... 263
Magnetism........... 261
Magneto-electricity,... 261
Magnets..... 261
Magnitude........... 25
Malleability. 28
Mariotte's Law........ 141
Measures, Standards of. 15
Mechanical Powers... 83
Mechanics, Principle of. 85
Microscopes........... 215
Mirage 202
Mirrors................ 191
Molecules............. 21
Molecular Forces... ... 35
Momentum............ 68
Motion............... 65
" Compound...... 71
" Circular....... 75
" in a Curve..... 79
" Laws of....... 70
" Perpetual....... 79
" Reflection of... 79
" Resistance to... 67
Music............ 163
Musical Scale..... ... 176

Near-sightedness. 222
Needle, Astatic. 303
" Magnetic....... 263
" Dipping...... 267
Newton's Rings........ 207
Nodal Lines.... 173
Nodes.. 171
Noise. 163

Northern Lights........ 282
Oceanic Currents...... 254
Octave................ 166
Opera-glass............ 218
Optics................ 185
Optical Instruments.... 215
Organ Pipes... 177
Oscillation, Centre of.. 60
Osmose of Gases...... 46
" Liquids....: 45
Overtones............ 175

Page's Rotating Machine.............. 305
Pendulum........... 58
Perpetual Motion 79
Pisa, Tower of........,.... 56
Pitch. 163
Platinum Wire......... 28
Pneumatics............ 132
Pneumatic Inkstand... 147
Polarization of Light... 210
" Heat... 225
" Electricity... 264
Porosity............... 21
Pressure of Air........ 132
" Gases...... 132
" Liquids. 103
Prince Rupert Drop.... 40
Prisms 199
Pulley................ 97
Pumps 142
" Air............ 132

Rain. 252
Rainbow.............. 212
Reaction............. 78
Reflected Motion...... 79
Relay. 309
Resonance....... 161
Rest................. 67
Ruhmkorff's Coil...... 312
Rupert Drop.......... 40

St. Elmo's Fire........ 284
Screw... 95
Sensitive Flame........ 182
Ship, Sailing of........ 74
Singing Flames....... 182
Siphon............... 145
Siren................. 164
Snow................. 253
Solution............. 44
Sonometer............. 170
Sound................ 151
" Intensity of... 157
" in a Vacuum... 155
" Interference of.. 169
" Reflection of.... 160

Sound, Refraction of... 156
" Superposition of 168
" Velocity of..... 155
Sounding-boards....... 175
Sound Waves........... 152
Speaking Tubes........ 158
" Trumpet 159
Specific Gravity....... 115
" " Flask.. 118
Spectrum, Solar....... 204
Spherical Aberration... 214
Spheroidal State....... 242
Steam................ 241
" -engine........... 246
Steelyard........... ... 88
Stereoscope 219
Stringed Instruments .. 170

Tacking............... 74
Telegraph..... 306
Telescope........... .. 216
Tempering... 39
Tenacity............. 21
Thermo-electricity...... 3.3
Thermometers.......... 236
Thunder.............. 2.4
Torsion Balance........ 31
Tourmaline.......... 211
Trade-wind........... 254
Turbine Wheel........ 126

Velocity................ 67
Vibrations of Air...... 152
" Cords..... 169
" Ether 189
" Pendulum 58
" Solids 228
" Sympathetic 178
Visual Angle............ 188
Voltaic Arch........... 296
" Battery...... 292
" Electricity.... 287
" Pair, The...... 289

Water.................. 255
" -barometer........ 142
" -level. 115
" -wheels.......... 125
Waves................. 128
Wave Motion........... 129
Wedge............. ... 96
Welding............... 36
Weight................. 49
Wells................. 112
Wheel and Axle........ 93
Wheel work............ 92
Wilde's Machine.... ... 311
Winds................ 253
Wind Instruments...... 178

www.ingramcontent.com/pod-product-compliance
Lightning Source LLC
Chambersburg PA
CBHW021155230426
43667CB00006B/410